BLUETOOTH® APPLICATION PROGRAMMING WITH THE JAVA™ APIs *ESSENTIALS EDITION*

The Morgan Kaufmann Series in Networking

Series Editor, David Clark, M.I.T.

For further information on these books and for a list of forthcoming titles, please visit our Web site at http://www.mkp.com.

BLUETOOTH® APPLICATION PROGRAMMING WITH THE JAVA™ APIs
ESSENTIALS EDITION

TIMOTHY J. THOMPSON

PAUL J. KLINE

C BALA KUMAR

MORGAN KAUFMANN PUBLISHERS

AN IMPRINT OF ELSEVIER SCIENCE

AMSTERDAM   BOSTON   LONDON   NEW YORK
OXFORD   PARIS   SAN DIEGO   SAN FRANCISCO
SINGAPORE   SYDNEY   TOKYO

Publishing Director:	Joanne Tracy
Publisher:	Denise E. M. Penrose
Acquisitions Editor:	Rick Adams
Publishing Services Manager:	George Morrison
Production Editor:	Lianne Hong
Assistant Editor:	Gregory Chalson
Design Direction:	Joanne Blank
Cover Design:	Dick Hannus
Cover Images:	iStockphoto
Composition:	Integra Software Services
Copyeditor:	Melissa Revell
Proofreader:	Phyllis Coyne et al. Proofreading Service
Indexer:	Keith Shostak
Interior Printer:	Sheridan Books, Inc.
Cover Printer:	Phoenix Color Corporation

Morgan Kaufmann Publishers is an imprint of Elsevier.
30 Corporate Drive, Suite 400, Burlington, MA 01803, USA

This book is printed on acid-free paper.

Library of Congress Cataloging-in-Publication Data
Thompson, Timothy J.
  Bluetooth application programming with the Java APIs/Tim J. Thompson, Paul J. Kline, and C Bala Kumar. – Essentials ed.
      p. cm. – (Morgan Kaufmann series in networking)
  C. Bala Kumar's name appeared first on t.p. of earlier ed.
  Includes bibliographical references and index.
    ISBN-13: 978-0-12-374342-8 (pbk. : alk. paper) 1. Bluetooth technology. 2. Java (Computer program language) 3. Application program interfaces (Computer software) 4. Wireless communication systems. I. Kline, Paul J. II. Kumar, C. Bala. III. Kumar, C. Bala. Bluetooth application programming with the Java APIs. IV. Title.
    TK5103.3.K86 2008
    004.6'2–dc22
                                        2007043858

ISBN: 978-0-12-374342-8

For information on all Morgan Kaufmann publications,
visit our Web site at www.mkp.com or www.books.elsevier.com

Printed in the United States of America
08 09 10 11 12 5 4 3 2 1

Working together to grow libraries in developing countries

www.elsevier.com | www.bookaid.org | www.sabre.org

ELSEVIER   BOOK AID International   Sabre Foundation

To my wife, Karmen, and son, Zane
—Tim

To my daughter, Rose, and her family, Terry, Morgan, and Andrew
—Paul

To my wife, Sundari, and sons, Sailesh and Shiva
—Bala

Contents

Preface

Bluetooth® wireless technology is a short-range radio standard that provides new opportunities for wireless devices. Originally, Bluetooth wireless technology was designed as a way of eliminating the cables attached to nearly all consumer electronic devices. However, the goals for Bluetooth wireless technology grew as its designers recognized that it enables a new kind of wireless network between electronic devices.

Since 2001, Java developers have had the opportunity to develop applications for a variety of wireless devices and cell phones. In 2000, the Java community recognized the importance of creating a standard extension to the Java programming language for use with Bluetooth devices. A standard application programming interface (API) for Bluetooth was needed because each Bluetooth software protocol stack had its own API for application programmers. These proprietary APIs meant that a Bluetooth application had to be ported to different Bluetooth stacks to run on different devices. Apart from the work involved in writing the code, interoperability testing on the various devices costs time and money for the involved companies. A standard API would help alleviate all these problems.

A team of experts from across the industry was assembled for this effort under Java Specification Request 82 (JSR-82). The result was a specification for Java APIs for Bluetooth wireless technology (JABWT). Since the release of JSR-82 in the spring of 2002, Bluetooth wireless technology has become a standard feature in cell phones with many of these phones also having support for JSR-82.

This book is based on the *Bluetooth Application Programming with the Java APIs* [2] written by the same authors. For this *Essentials Edition*, the authors have updated the background information to reflect the changes that have occurred in the area of Bluetooth wireless technology and

JSR-82, including support for MIDP Push, since *Bluetooth Application Programming with the Java APIs* was published in 2004. While reading the JSR-82 specification document provides you with a description of the API, this book provides you with the rationale and best practices in utilizing the API.

The objectives of this book are to

Give an overview of Java™ Platform, Micro Edition (Java ME) and Bluetooth wireless technology

Outline the JABWT architecture

Explain the API in detail

Intended Audience

The book is intended for software developers, academics, and other professionals who want to develop Java software for Bluetooth devices. To gain the most out of this book, you will find it helpful to have a working knowledge of Java ME and familiarity with Bluetooth wireless technology. The book cites several references that provide additional information on these subjects. We believe that a Java ME programmer will need no additional reading beyond this book to write JABWT applications.

If you would like more examples or more information on developing and porting JSR-82 to a handset, the authors recommend the predecessor to this book: *Bluetooth Application Programming with the Java APIs*.

How This Book Is Organized

Different readers of this book will be seeking different information. We have identified three sets of people:

1. Those looking for an overview to make decisions on projects
2. Those who will be leading projects or managing projects in this area
3. Programmers who need detailed information on how to program using JABWT

Apart from the introductory chapters, the chapters are organized into three main sections to accommodate the three sets of people identified above. The three divisions are

1. Overview: The executive introduction
2. API capabilities: The explanation for the project manager
3. Programming with the API: The programmer's guide

Readers can choose the sections that suit their needs in each chapter. Chapters 1 through 3 are overview chapters. Chapters 4 through 9 detail the various sections of the API. Chapter 9 describes the MIDP Push capabilities added since the last book. Throughout the book many code examples are given to explain the API. The complete JSR-82 API is available at www.jcp.org/en/jsr/detail?id=82.

There is a website for this book where you can access the complete code examples found in the book. In addition, you can find the latest news about JABWT, book errata, and other useful links. To access the website, go to www.mkp.com and use the search option with the title of this book.

The topics in this book are organized as follows:

Chapter 1, Introduction, presents an overview of Bluetooth wireless technology and Java ME. It also provides a context for the JABWT specification.

Chapter 2, An Overview of JABWT, defines the goals, characteristics, and scope of JABWT.

Chapter 3, High-Level Architecture, presents the high-level architecture of JABWT.

Chapter 4, RFCOMM, discusses the APIs for Bluetooth serial port communications using RFCOMM.

Chapter 5, OBEX, introduces the architecture and the APIs for making OBEX connections.

Chapter 6, Device Discovery, discusses the APIs for Bluetooth device discovery.

Chapter 7, Service Discovery, describes the APIs for service discovery and service registration.

Chapter 8, L2CAP, presents the API for Bluetooth communications using the logical link control and adaptation protocol.

Chapter 9, Push Registry, describes the support available in JABWT for the Push Registry as described in MIDP 2.0.

About the Authors

Timothy J. Thompson is a Principal Software Engineer on the Advanced Technology and Architecture team in Motorola's Mobile Device Business. He is currently the JSR-82 Maintenance Lead. He was the OBEX architect on the JABWT specification team at Motorola. He received his Master's degree in Computer Science from Texas A&M University.

Paul J. Kline manages a team that develops Linux Board Support Packages for the Multimedia Applications Division at Freescale Semiconductor. Previously, he worked at Motorola where he was a member of the JSR-82 Expert Group and the first JSR-82 Maintenance Lead. He received his Ph.D. in Mathematical Psychology from the University of Michigan.

C Bala Kumar manages platform development for the Multimedia Applications Division at Freescale Semiconductor. Previously, he worked at Motorola where he chaired the industry expert group that defined the Java APIs for Bluetooth wireless technology. He received his Master's degree in Electrical Engineering from the University of Texas at Austin.

Acknowledgments

A large number of people were involved with the original development of the Java APIs for Bluetooth wireless technology. As the three of us set out to write a book explaining those Bluetooth APIs, we were pleased to discover that we would again receive contributions and assistance from a large number of dedicated and talented individuals.

The authors thank Glade Diviney, Peter Kembro, and Ashwin Kamal Whitchurch for reviewing the book *Bluetooth Application Programming with the Java APIs*, which is the basis of this book, and making valuable comments and suggestions. Thanks also to R. Thiagarajan, N. Murugan, Franck Thibaut, Ramesh Errabolu, Ranjani Vaidyanathan, and Ravi Viswanathan, who commented on various chapters of the original book. Of course, the authors are totally responsible for any errors that remain.

When the original book was in the proposal stage, we received excellent advice and suggestions from Alok Goyal, Teck Yang Lee, Girish Managoli, Brent Miller, Venugopal Mruthyunjaya, N. Ramachandran, Rajeev Shorey, and Mark Vandenbrink. Ashwin Whitchurch, Brent Miller, Glade Diviney, and Girish Managoli provided additional feedback on the proposal for this book.

The Java APIs for Bluetooth wireless technology were developed by a team of industry experts, the JSR-82 expert group, and the team at Motorola that drafted the specification, wrote the reference implementation, and developed the conformance tests. The authors believe that the efforts and contributions of all these individuals produced an API that will have important benefits to the Java community. The authors would like to thank the members of the JSR-82 expert group for all their work on the API: Jouni Ahokas, Patrick Connolly, Glade Diviney, Masahiro Kuroda, Teck Yang Lee, Paul Mackay, Brent Miller, Jim Panian,

Farooq Anjum, Charatpong Chotigavanich, Peter Dawson, Peter Duchemin, Jean-Philippe Galvan, Daryl Hlasny, Knud Steven Knudsen, Andrew Leszczynski, Martin Mellody, Anthony Scian, and Brad Threatt.

We greatly appreciate all of the contributions of the other members of the JSR-82 team at Motorola: Lawrence Chan, Judy Ho, Will Holcomb, Judy Lin, Mitra Mechanic, Ramesh Errabolu, Ranjani Vaidyanathan, Ravi Viswanathan, and Allen Peloquin. Jim Erwin, Jim Lynch, Aler Krishnan, Ed Wiencek, and Mark Patrick provided a great deal of assistance to the JSR-82 team.

We would also like to thank Khurram Qureshi and Mark Vandenbrink for their help and support in making this book a reality.

The authors are very grateful to Rick Adams, Gregory Chalson, Lianne Hong, Karyn Johnson, and Mamata Reddy of Morgan Kaufmann for all their hard work

Tim thanks his wife, Karmen, for her encouragement, patience, and support.

Paul thanks his wife, Dianne, for her support and encouragement.

Bala thanks Sundari, Sailesh, and Shiva for their understanding and support through long nights and weekends working on this project. Bala also thanks his mother, Suseela, and sister, Surya, for all their patient nurturing and Mr. B. Kanakasabai for being his lifelong friend and mentor.

Tim Thompson

Paul Kline

C Bala Kumar

1 Introduction

This chapter begins with an introduction to wireless connectivity and Bluetooth® wireless technology. It then gives

- An overview of the Bluetooth protocol stack
- An overview of the Java Platform, Micro Edition
- A description of the need for Java technology in Bluetooth devices

1.1 Wireless Connectivity

We are in the information age. The term "information age" came about because of the exchange of massive amounts of data between computing devices using wired and wireless forms of communication. We are rapidly moving toward a world in which communications and computing are ubiquitous.

Today, high-speed networks connect servers, personal computers, and other personal computing devices. High-end routers manage the networks. The distinction between voice and data networks has blurred, and the same network tends to carry both types of traffic. The desire and need to communicate with distant computers led to the creation of the Internet. The days of consumers buying a personal computer for stand-alone applications have disappeared. These days the primary motive for buying a personal computer is to use it as a communication tool so that one can have Internet access to e-mail and the World Wide Web. The same is true of today's embedded computing devices. Instead of simply being an organizer or phone, embedded computing devices have become another way to access the Internet.

Increased dependence on the Internet and the need to stay connected from anywhere at all times have led to advances in mobile computing and communications. We have been communicating without wires for some time with satellites, cordless phones, cellular phones, and remote-control devices. However, in recent years the wireless communications industry has seen explosive growth. Long-range wireless communication invariably uses radio frequency (RF). Typically, long-range communications use the licensed parts of the RF spectrum, and user fees apply. Short-range communications can use either RF or infrared and typically use unlicensed (i.e., free) parts of the frequency spectrum.

There are many short-range wireless standards, but the three main ones are Infrared from the Infrared Data Association® (IrDA®), Bluetooth wireless technology, and wireless local area network (WLAN). WLAN is also known as IEEE 802.11, and it comes in several variants (802.11b, 802.11g, 802.11a, 802.11n, etc.), which operate at 2.4 gigahertz (GHz) or 5 GHz. The IrDA created a wireless communications system that makes use of infrared light. Whereas RF communication can penetrate many objects, IrDA is limited to line of sight. Both 802.11b and Bluetooth wireless technologies communicate in the 2.4-GHz RF band but are aimed at different market segments. The 802.11 technology has a longer range but consumes substantially more power than Bluetooth wireless technology. The 802.11 variant is primarily for data. The only protocol for supporting voice is Voice over Internet Protocol (VoIP). Table 1.1 provides a comparison of these three technologies.

Wireless communications allow computing and communication devices to be used almost anywhere and to be used in new, progressive ways. The increase in wireless mobile Internet devices is proof that wireless connectivity is pervasive. Powerful software programming environments will help fuel this mobile computing explosion by enabling the development of compelling applications. The Java platform provides a powerful programming environment that has great promise for wireless devices. Many mobile devices now come with support for Java Platform, Micro Edition (Java ME) programs. This book explains how to program Bluetooth applications with the Java programming language.

Table 1.1 Comparison of Wireless Communication Technologies

Feature and function	IrDA	Wireless LAN	Bluetooth communication
Connection type	Infrared, narrow beam, line of sight	Spread spectrum, spherical	Spread spectrum, spherical
Spectrum	Optical 850–900 nm	RF 2.4 GHz (5 GHz for 802.11a/n)	RF 2.4 GHz
Transmission power	40–500 mW/Sr	100 mW	10–100 mW
Maximum data rate	9600 bps–16 Mbps (very rare)	11 Mbps (54 Mbps for 802.11a, 802.11g)	3 Mbps
Range	1 m	100 m	10–100 m
Supported devices	2	Connects through an access point	8 (active), 200 (passive)
Voice channels	No	No	Yes
Addressing	32-bit physical ID	48-bit MAC	48-bit MAC

1.2 What Is Bluetooth Wireless Technology?

Bluetooth wireless technology is an open specification for a low-cost, low-power, short-range radio technology for ad hoc wireless communication of voice and data anywhere in the world. Let us examine each of these attributes.

- An open specification means that the specification is publicly available and royalty free.
- Short-range radio technology means devices can communicate over the air using radio waves at a distance of 10 meters (m). With higher transmission power the range increases to approximately 100 m.
- Because communication is within a short range, the radios are low power and are suited for portable, battery-operated devices.
- Bluetooth wireless technology supports both voice and data, allowing devices to communicate either type of content.
- Bluetooth wireless technology works anywhere in the world because it operates at 2.4 GHz in the globally available, license-free, industrial, scientific, and medical (ISM) band.

Because the ISM frequency band is available for general use by ISM applications, several other devices (e.g., WLAN, cordless phones, microwave ovens) operate in this band. Bluetooth wireless technology is designed to be very robust in the face of interference from other devices.

1.2.1   History of Bluetooth Wireless Technology

Bluetooth communications originated in 1994, when Ericsson began a study to find alternatives to connecting mobile phones to its accessories. The engineers looked at a low-power and low-cost radio interface to eliminate cables between the devices. However, the engineers also realized that for the technology to be successful, it had to be an open standard, not a proprietary one. In early 1998, Ericsson joined Intel, International Business Machines (IBM), Nokia, and Toshiba and formed the Bluetooth Special Interest Group (SIG) to focus on developing an open specification for Bluetooth wireless technology. The original companies, known as promoter companies, publicly announced the global Bluetooth SIG in May 1998 and invited other companies to join the Bluetooth SIG as Bluetooth adopters in return for a commitment to support the Bluetooth specification. In July 1999, the Bluetooth SIG published version 1.0 of the Bluetooth specification. In December 1999, four new promoter companies—3Com, Agere, Microsoft, and Motorola—joined the Bluetooth SIG (Figure 1.1).

Since then, the awareness of Bluetooth wireless technology has increased, and many other companies have joined the Bluetooth SIG as adopters, which gives them a royalty-free license to produce Bluetooth-

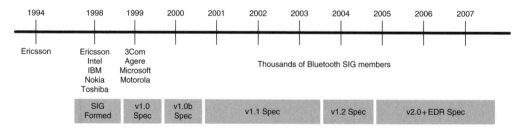

Figure 1.1   Bluetooth SIG time line.

enabled products. Adopter companies also have early access to specifications and the ability to comment on them. Interest in the Bluetooth SIG has grown, and there are currently thousands of member companies. These companies represent academia and a variety of industries.

Why is this technology called Bluetooth wireless technology? It was named after a Danish Viking king, Harald Blåtand, who ruled circa A.D. 940–981. *Blåtand* loosely translates to "blue tooth." During his reign, King Harald Blåtand is supposed to have united and controlled Denmark and Norway. Because this new radio technology was expected to unify the telecommunication and computing industries, it seemed fitting to name it after King Harald. A part-time historian on the team proposed *Bluetooth* as the internal code name. Because the Bluetooth SIG marketing team could not come up with a better name that was not already trademarked, the name stuck.

1.2.2   Bluetooth Vision

Bluetooth wireless technology was originally developed as a cable replacement technology for connecting devices such as mobile phone handsets, headsets, and portable computers with each other (Figure 1.2). However, wireless connectivity between fixed and mobile devices enables many usage scenarios other than cable replacement. By enabling wireless links and communication between devices, a short-range wireless network was created that gave rise to the notion of a personal area network. Designed as an inexpensive wireless networking system for all classes of portable devices, Bluetooth devices have the capability to form ad hoc networks. These networks should enable easy and convenient connections to printers, Internet access points, and personal devices at work and at home.

There are many usage scenarios for Bluetooth wireless technology, and the technology has been put to wide use. Let us look at a couple of the usage models.

The three-in-one phone usage model allows a mobile telephone to be used as a cellular phone in the normal manner, as a cordless phone that connects to a voice access point (e.g., cordless base station), and as an intercom or "walkie-talkie" for direct communication with another device. The cordless telephony and the intercom features use Bluetooth wireless technology.

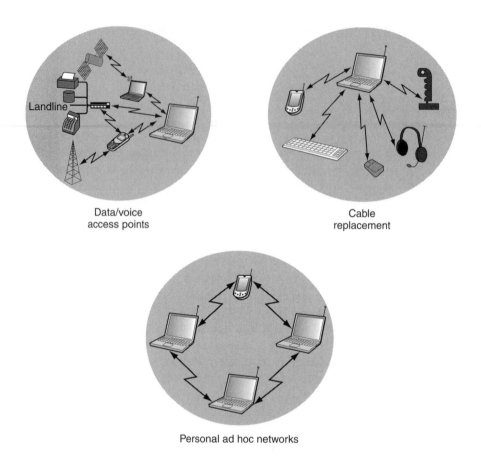

Data/voice
access points

Cable
replacement

Personal ad hoc networks

Figure 1.2   Bluetooth use cases.

The second use case is wireless telematics. Assume that a user who is talking on a cell phone approaches his or her automobile but wants to continue the phone conversation in the hands-free mode. Using Bluetooth communication the user can continue the phone conversation using the microphone and speakers equipped in the dashboard of the automobile.

Another use case is the instant postcard, whereby a user with a digital camera transmits a photo via a data access point, which could be a mobile phone. Similar use cases include automatic synchronization, business card exchange, hotel and airline check-in, electronic ticketing, and wireless games.

1.2.3 Bluetooth Specification

The Bluetooth specification is the result of cooperation by many companies under the Bluetooth SIG umbrella. The specification defines over-the-air behavior to ensure the compatibility of Bluetooth devices from different vendors. It defines the complete system from the radio up to the application level, including the software stack. The specification is very lengthy because of the breadth of topics it covers. At the highest level, the specification is really a collection of specifications: a Bluetooth Core Specification, Bluetooth Transport Specifications, Bluetooth Protocol Specifications, and Bluetooth Profile Specifications. The Bluetooth Core Specification defines the overall Bluetooth architecture, common terms used within the community, a controller specification, and a host specification. Bluetooth Transport Specifications describe how different transports can be used to communicate between the host and the controller (e.g., UART, USB, Three-Wire). Bluetooth Protocol Specifications describe higher level protocols that run over Bluetooth Wireless Technology (e.g., RFCOMM and OBEX). Finally, Bluetooth Profile Specifications are individual specifications that describe individual Bluetooth profiles.

Bluetooth profiles, essentially usage models, describe how applications are to use the Bluetooth stack. A Bluetooth profile is a set of capabilities of the protocol layers that represent a default solution for a usage model. Bluetooth profiles are the basis of Bluetooth protocol stack qualification, and any new implementations of a Bluetooth profile have to go through the qualification process described herein. The specification and profiles continue to evolve as new areas are identified in which Bluetooth wireless technology can be used. Bluetooth protocols and profiles are discussed briefly in the next section.

1.3 Overview of Bluetooth Stack Architecture

This section provides a brief overview of the Bluetooth protocol stack. The Bluetooth protocol stack can be broadly divided into two components: the

Bluetooth host and the Bluetooth controller (or Bluetooth radio module). The Host Controller Interface (HCI) provides a standardized interface between the Bluetooth host and the Bluetooth controller. Figure 1.3 illustrates the Bluetooth host and Bluetooth device classification.

The Bluetooth host is also known as the upper-layer stack and usually is implemented in software. It is generally integrated with the system software or host operating system. Bluetooth profiles are built on top of the protocols. They are generally in software and run on the host device hardware. For example, a laptop computer or a phone would be the host device. The Bluetooth host would be integrated with the operating system on the laptop or the phone.

The Bluetooth radio module or controller usually is a hardware module like a PC card (see Figure 1.3) that plugs into a target device. More and more devices have the Bluetooth controller built into the device. The upper stack interfaces to the Bluetooth radio module via the HCI. The Bluetooth radio module usually interfaces with the host system via one of the standard input/output (I/O) mechanisms, such as the universal asynchronous receiver-transmitter (UART) and universal serial bus (USB). Although the Bluetooth host and the Bluetooth controller classifications apply to most devices, the two are integrated in some devices—headsets, for example—and HCI is not used. The various blocks shown in Figure 1.3 are part of the Bluetooth protocol stack, which is discussed next.

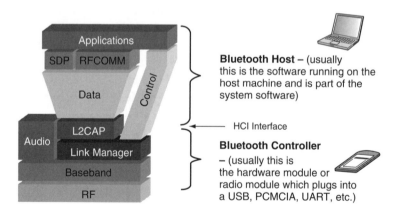

Figure 1.3   Bluetooth host and device classification.

1.3.1 Bluetooth Protocols

Figure 1.4 shows a block diagram of the Bluetooth protocol stack. Several protocols are defined in the Bluetooth specification, but Figure 1.4 shows the common ones. Shaded boxes represent the protocols addressed by Java APIs for Bluetooth wireless technology (JABWT, where API stands for application programming interface). The protocol stack is composed of protocols specific to Bluetooth wireless technology, such as the Service Discovery Protocol (SDP), and other adopted protocols, such as the Object Exchange protocol (OBEX™).

- The Bluetooth radio (layer) is the lowest defined layer of the Bluetooth specification. It defines the requirements of the Bluetooth transceiver device operating in the 2.4-GHz ISM band.

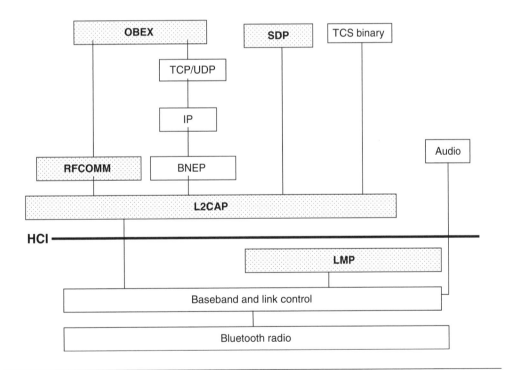

Figure 1.4   Bluetooth protocol stack.

- The baseband and link control layer enables the physical RF link between Bluetooth units making a connection. The baseband handles channel processing and timing, and the link control handles the channel access control. There are two different kinds of physical links: synchronous connection oriented (SCO) and asynchronous connectionless (ACL). An ACL link carries data packets, whereas an SCO link supports real-time audio traffic.

- Audio is really not a layer of the protocol stack, but it is shown here because it is uniquely treated in Bluetooth communication. Audio data is typically routed directly to and from the baseband layer over an SCO link. Of course, if a data channel is used (e.g., in VoIP applications), audio data will be transmitted over an ACL link.

- The Link Manager Protocol (LMP) is responsible for link setup and link configuration between Bluetooth devices, managing and negotiating the baseband packet sizes. The LMP manages the security aspects, such as authentication and encryption, by generating, exchanging, and checking link and encryption keys.

- The HCI provides a command interface to the radio, baseband controller, and link manager. It is a single standard interface for accessing the Bluetooth baseband capabilities, the hardware status, and control registers.

- The Logical Link Control and Adaptation Protocol (L2CAP) shields the upper-layer protocols from the details of the lower-layer protocols. It multiplexes between the various logical connections made by the upper layers.

- The SDP provides a means for applications to query services and characteristics of services. Unlike in an LAN connection, in which one connects to a network and then finds devices, in a Bluetooth environment one finds the devices before one finds the service. In addition, the set of services available changes in an environment when devices are in motion. Hence SDP is quite different from service discovery in traditional network-based environments. SDP is built on top of L2CAP.

- Serial ports are one of the most common communications interfaces used in computing and communication devices.

The RFCOMM protocol provides emulation of serial ports over L2CAP. RFCOMM provides transport capabilities for upper-level services that use a serial interface as a transport mechanism. RFCOMM provides multiple concurrent connections to one device and provides connections to multiple devices.

- Bluetooth-enabled devices will have the ability to form networks and exchange information. For these devices to interoperate and exchange information, a common packet format must be defined to encapsulate layer 3 network protocols. The Bluetooth Network Encapsulation Protocol (BNEP) [5] is an optional protocol that encapsulates packets from various networking protocols. The packets are transported directly over L2CAP.

- Telephony Control Protocol Specification, Binary (TCS binary) defines the call control signaling for establishment of voice and data calls between Bluetooth devices. It is built on L2CAP.

- Adopted protocols, such as OBEX and the Internet Protocol (IP), are built on one of the protocols discussed earlier (e.g., OBEX is built on RFCOMM, and IP is built on BNEP).

- The Bluetooth SIG also is defining newer protocols built on one of the protocols discussed earlier, but mainly they are built on top of L2CAP. The Audio/Video Control Transport Protocol [7] and Audio/Video Distribution Transport Protocol [8] are examples of some newer protocols.

1.3.2 Bluetooth Profiles

In addition to the protocols, Bluetooth profiles have been defined by the Bluetooth SIG [1]. A Bluetooth profile defines standard ways of using selected protocols and protocol features that enable a particular usage model. In other words, it defines how different parts of the Bluetooth specification can be used for a particular use case. A profile can be described as a vertical slice through the protocol stack. It defines options in each protocol that are needed for the profile. The dependency of the profiles on protocol layers and features varies. Two profiles may use a different set of protocol layers and a different set of features within the same protocol layer.

A Bluetooth device can support one or more profiles. The four "basic" profiles are the Generic Access Profile (GAP) [1], the Serial Port Profile (SPP) [9], the Service Discovery Application Profile (SDAP) [10], and the Generic Object Exchange Profile (GOEP) [11].

- The GAP is the basis of all other profiles. Strictly speaking, all profiles are based on the GAP. GAP defines the generic procedures related to establishing connections between two devices, including the discovery of Bluetooth devices, link management and configuration, and procedures related to the use of different security levels.

- The SDAP describes the fundamental operations necessary for service discovery. This profile defines the protocols and procedures to be used by applications to locate services in other Bluetooth-enabled devices.

- The SPP defines the requirements for Bluetooth devices necessary for setting up emulated serial cable connections using RFCOMM between two peer devices. SPP maps to the RFCOMM protocol directly and enables legacy applications using Bluetooth wireless technology as a cable replacement.

- The GOEP is an abstract profile on which concrete usage case profiles can be built. These are profiles using OBEX. The profile defines all elements necessary for support of the OBEX usage models (e.g., file transfer, synchronization, or object push).

Figure 1.5 shows the relationships among the various Bluetooth profiles. The Bluetooth profiles are hierarchical. For example, the File Transfer Profile is built on top of GOEP, which depends on SPP, which is built upon GAP. Bluetooth profiles also can be classified on the basis of the functional or services point of view. From a programming perspective, however, it is the profile hierarchy that is applicable. The basic profiles— GAP, SDAP, SPP, and GOEP—also are known as *transport profiles*, upon which other profiles, known as *application profiles*, can be built.

Many profiles are based on the basic profiles. Figure 1.5 will probably be obsolete soon because more profiles are being developed continuously. Refer to the Bluetooth Web site (www.bluetooth.com) for specifications on the latest available profiles.

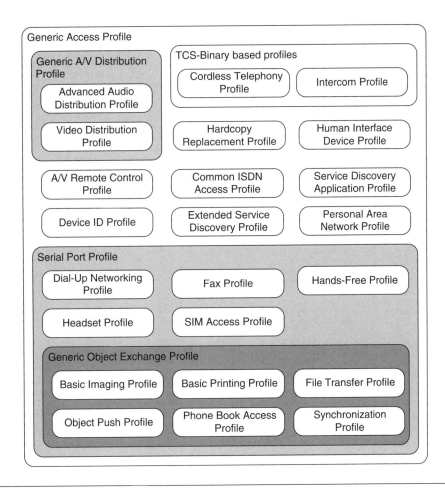

Figure 1.5 Bluetooth profile hierarchy.

1.3.3 Bluetooth Qualification

Bluetooth qualification is the certification process required for any product using Bluetooth wireless technology. The qualification process ensures that products comply with the Bluetooth specification. Only qualified products are entitled to use the free license to the patents required to implement Bluetooth wireless technology, the Bluetooth brand, and the Bluetooth logo. Essentially, there are three levels of Bluetooth qualification testing:

- Core specification conformance
- Interoperability testing to ensure that devices work with one another at the profile level
- Bluetooth branding conformance

More details on the qualification process can be obtained from the Bluetooth Qualification Program Web site [12].

1.4   What Is JAVA ME?

This section gives a brief overview of Java ME (formerly known as J2ME). For details about Java ME, refer to books by Topley [13] and Riggs and colleagues [14].

Java ME is the Java platform for consumer and embedded devices such as mobile phones, pagers, personal organizers, television set-top boxes, automobile entertainment and navigation systems, Internet televisions, and Internet-enabled phones. Java ME is one of the three platform editions. The other two platform editions are Java Platform, Enterprise Edition (Java EE) for servers and enterprise computers and Java Platform, Standard Edition (Java SE) for desktop computers. A related technology is Java Card™ technology. The Java Card specification enables Java technology to run on smart cards and other devices with more limited memory than a low-end mobile phone. These groupings are needed to tailor the Java technology to different areas of today's vast computing industry. Figure 1.6 illustrates the Java platform editions and their target markets.

The Java ME platform brings the power and benefits of Java technology (code portability, object-oriented programming, and a rapid development cycle) to consumer and embedded devices. The main goal of Java ME is to enable devices to dynamically download applications that leverage the native capabilities of each device. The Consumer and embedded space covers a range of devices from pagers to television set-top boxes that vary widely in memory, processing power, and I/O capabilities. To address this diversity, the Java ME architecture defines

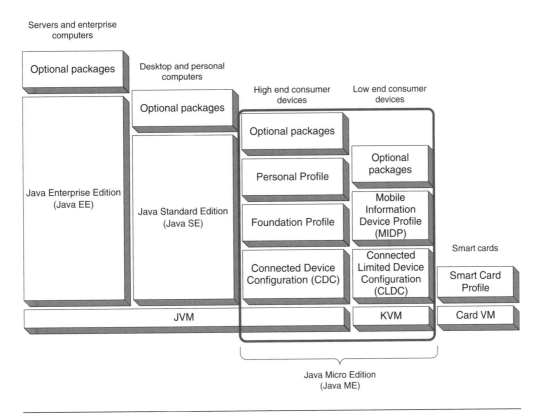

Figure 1.6   Java platforms.

configurations, profiles, and optional packages to allow for modularity and customizability. Figure 1.7 shows the high-level relationship between layers of the Java ME architecture. The layers are explained further in the next section.

1.4.1   Configurations

A Java virtual machine interprets the Java byte codes generated when Java programs are compiled. A Java program can be run on any device that has a suitable virtual machine and a suitable set of Java class libraries.

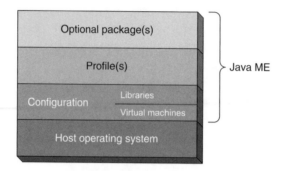

Figure 1.7   Components of Java ME architecture.

Configurations are composed of a Java virtual machine and a mini-mal set of class libraries. The Java virtual machine usually runs on top of a host operating system that is part of the target device's system software. The configuration defines the minimum functionality for a particular category or grouping of devices. It defines the minimum capabilities and requirements for a Java virtual machine and class libraries available on all devices of the same category or grouping. Currently, there are two Java ME configurations: the Connected, Limited Device Configuration (CLDC) [25] and the Connected Device Configuration (CDC) [26].

Connected, Limited Device Configuration

The CLDC focuses on low-end consumer devices and is the smaller of the two configurations. Typical CLDC devices, such as personal organizers, mobile phones, and pagers, have slow processors and limited memory, operate on batteries, and have only intermittent network connections. A CLDC implementation generally includes a kilobyte virtual machine (KVM). It gets its name because of its small memory footprint (on the order of kilobytes). The KVM is specially designed for memory-constrained devices.

Connected Device Configuration

The CDC focuses on high-end consumer devices that have more mem-ory, faster processors, and greater network bandwidth. Typical examples of CDC devices are television set-top boxes and high-end communicators. CDC includes a virtual machine that conforms fully to the Java Virtual Machine Specification [17]. CDC also includes a much larger subset of the Java SE platform than CLDC.

1.4.2 Profiles

Configurations do not usually provide a complete solution. Profiles add the functionality and the APIs required to complete a fully functional run-time environment for a class of devices. Configurations must be combined with profiles that define the higher level APIs for providing the capabilities for a specific market or industry. It is possible for a single device to support several profiles. Examples of profiles are Mobile Information Device Profile (MIDP), Foundation Profile (FP), and Personal Profile (PP). A clarification is needed: The Bluetooth profiles defined previously are not to be confused with the Java ME profiles discussed here. The two profiles are not related. A *Bluetooth profile* refers to a set of functionality of the Bluetooth protocols for a particular usage case. Java ME profiles are a set of APIs that extend the functionality of a Java ME configuration.

Mobile Information Device Profile

The first profile that was created was MIDP [27]. This profile is designed for mobile phones, pagers, and entry-level personal organizers. MIDP combined with CLDC offers core application functionality, such as a user interface, network capability, and persistent storage. MIDP provides a complete Java run-time environment for mobile information devices. MIDP applications are called *MIDlets*. MIDlet is a class defined in MIDP and is the superclass for all MIDP applications.

Foundation Profile

The FP [19, 28] is the lowest level profile for CDC. Other profiles can be added on top as needed to provide application functionality. The FP is meant for embedded devices without a user interface but with network capability.

Personal Profile

The PP [20, 29] is for devices such as high-end personal organizers, communicators, and game consoles that require a user interface and Internet applet support. PP replaces PersonalJava™ technology and provides PersonalJava applications a clear migration path to the Java ME platform.

In addition there is a Personal Basis Profile (PBP) [21, 30], which is a subset of PP aimed at devices that require only a basic level of graphical presentation (e.g., television set-top boxes).

1.4.3   Optional Packages

Many Java ME devices include additional technologies, such as Bluetooth wireless technology, multimedia, wireless messaging, and database connectivity. Optional packages were created to fully leverage these technologies through standard Java APIs. Device manufacturers can include these optional packages as needed to fully utilize the features of each device.

In addition to the configurations, profiles, and optional packages, device manufacturers are able to define additional Java classes to take advantage of features specific to the device. These classes are called *licensee open classes* (LOCs). An LOC defines classes available to all developers. *Licensee closed classes* define classes available only to the device manufacturer. Programs using these classes may not be portable across devices having the same configuration and profiles.

1.5   Why JAVA Technology for Bluetooth Devices?

How an end user uses Bluetooth wireless technology varies from person to person. Two people with the same model of a Bluetooth-enabled phone might want to use it for different purposes. One person might want to be able to download video games to the phone and use the phone as a television remote control. The other person might want to use the same model phone to unlock car doors, operate kitchen appliances, and open and close garage doors. One way for both people to achieve their goals is to make it possible to download Bluetooth applications onto personal organizers and mobile phones to customize those handheld devices. To make downloading applications a reality, one needs a standard API that lets programmers write Bluetooth applications that work across many hardware platforms. To define this standard API, the Java language is the ideal choice. A Java API enables applications to run on different types of hardware, operating systems, and classes of device. In addition to portability, the Java language provides several other benefits:

- Rapid development of applications because of better abstractions and high-level programming constructs provided by an object-oriented programming language.
- Ability to dynamically expand a program's functionality during execution by loading classes at run time.
- Class file verification and security features that provide protection against malicious applications. These safeguards are required to customize devices by downloading applications.
- Standards with better user interfaces that support sophisticated user interaction.
- Large developer community. The number of people who program in the Java language is continuously growing. The developer talent needed for programming in the Java language already exists, and there is no need to grow a developer community.

For these reasons, the decision was made to develop a standard API for Bluetooth wireless technology using the Java programming language. This standardization effort resulted in JABWT. As shown later in this book, this standardization effort complements existing technologies rather than replacing them. JABWT is built on top of the already established and widely used Bluetooth protocol stack.

1.5.1 Java Community Process^{SM} (JCP) and JSR-82

Standard APIs in the Java programming language are defined though the JCP. The JCP coordinates the evolution of the Java programming language. Each new API is developed as a Java Specification Request

(JSR). All Java ME configurations, profiles, and optional packages are defined as JSRs. The process for defining a standard Java API is as follows:

1. The potential specification lead submits a new JSR.
2. The JCP executive committee reviews and votes on the JSR.
3. After JSR approval, the specification lead forms an expert group.
4. The expert group defines the specification.
5. JCP members review the specification during the community review period.
6. The specification is open for public review.
7. The specification lead submits the specification as the proposed final draft.
8. The executive committee votes on the specification to accept or deny the API.
9. If the vote passes, the final release of the specification is announced.

The process just described was followed in standardizing the JABWT under JSR-82 [22]. The expert group that defined JABWT consisted of 18 companies and three individuals. The companies were Extended Systems, IBM, Mitsubishi Electric, Motorola, Newbury Networks, Nokia, Parthus Technologies, Research in Motion, Rococo Software, Sharp Laboratories of America, Sony Ericsson Mobile Communications, Smart Fusion, Smart Network Devices, Sun Microsystems, Symbian, Telecordia, Vaultus, and Zucotto. The API was defined as an optional package for Java ME devices based on CLDC.

1.5.2   What about Java SE?

Because Bluetooth wireless technology can be found in Java SE devices, you may ask why this standardization effort focused on Java ME devices. The expert group believed that the initial set of devices that would use Java language capabilities over Bluetooth protocols would be in the Java ME device space. However, as the next chapters show, the API was defined in such a way as to rely heavily on one set of CLDC APIs known as the *Generic Connection Framework* (GCF).

That thinking partially paid off. JSR-197 (Generic Connection Framework Optional Package) [23] defined the GCF for Java SE. Those Java SE platforms that include JSR-197 may support JABWT in the future.

1.6 Summary

The JABWT specification provides a standard set of APIs for developing Bluetooth applications. The Java APIs defined by JABWT are considered optional packages for Java ME. Applications written with JABWT are potentially portable to a wide range of devices with a wide range of Bluetooth radio modules and Bluetooth protocol stacks.

This chapter gave an overview of Bluetooth wireless technology and Java ME. These are two very large topics. To learn more about Bluetooth wireless technology, refer to the Bluetooth specifications [1] or books on the subject [3, 4]. The following Web sites are good places to start:

www.bluetooth.com

www.palowireless.com

To learn more about Java ME, see the books by Topley [13] and by Riggs and associates [14]. Helpful Web sites for Java ME and JABWT are

java.sun.com

www.jcp.org

www.jcp.org/jsr/detail/82.jsp

There are several articles [24], white papers, and tutorials on these subjects on the Web. There are several newsgroups on Bluetooth wireless technology, but the following is devoted to JABWT specifically:

groups.yahoo.com/group/jabwt

This chapter noted the need for Java technology on Bluetooth devices and explained the process of defining JABWT.

2 An Overview of JABWT

CHAPTER

This chapter describes:

- The goals of the JABWT specification
- The characteristics of the JABWT specification
- The scope of the JABWT specification

Some sections in the chapter may not seem relevant for those primarily interested in programming with JABWT. But the overview of JABWT is presented to lead to a better understanding of the capabilities and the reasoning behind these APIs.

2.1 Goals

The Bluetooth specification defines the over-the-air behavior for ensuring compatibility of Bluetooth devices from different vendors. The Bluetooth specification does not standardize a software API to Bluetooth stacks for use by Bluetooth applications. JABWT helps solve this problem by defining the first standard API for Bluetooth application developers. The overall goal of the JABWT standardization effort discussed in this book is to define a standard set of APIs that will enable an open, third-party application development environment for Bluetooth wireless technology.

The goals were to minimize the number of classes (the total number of classes in JABWT is 21); keep the API simple and easy to learn and program; and keep it powerful. The meaningful high-level abstractions help in third-party application development. This API brings together the benefits of two different technologies: Bluetooth wireless technology and Java technology. Having this standard API in the Java language brings in all the benefits of Java technology, some of which are discussed in Chapter 1. The abstractions and ease of programming of the Java language facilitate easy development of complex programs. The goal of JABWT is to present access to Bluetooth wireless technology in the easy but powerful form of the Java language.

2.1.1   Target Devices

JABWT is aimed mainly at devices that are limited in processing power and memory and are primarily battery operated. These devices can be manufactured in large quantities. Low cost and low power consumption are primary goals of the manufacturers. JABWT takes these factors into consideration. Figure 2.1 shows the types of devices that might use JABWT. Some of the devices shown, such as the car, laptop, and LAN access point, are not Java ME devices. These devices are likely to operate with Java SE or CDC. Some manufacturers of these products, however, are already incorporating JABWT in their designs. In addition, work completed under JSR-197 will make integrating JABWT into these products easier. JSR-197 [23] created an optional package out of GCF alone, allowing applications that rely on the GCF to migrate to Java SE. JSR-197 also is intended to use GCF APIs as defined by the Java ME Foundation Profile along with improvements proposed in CLDC 1.1 (JSR-139) [25].

2.1.2   Keeping up with the Bluetooth Profiles

One initial idea was to define an API based on the Bluetooth profiles. But the JSR-82 expert group realized that the number of Bluetooth profiles is constantly growing and that it would not be possible to keep up with the new profiles in the JABWT specification. Instead the JSR-82 expert group decided to provide support for only basic protocols and profiles rather than introducing new API elements for each Bluetooth profile. The intent of the JABWT design is to enable new Bluetooth profiles to be built on top of this API with the Java programming language. Bluetooth profiles are being built on top of OBEX, RFCOMM, and L2CAP. For this reason, all three of these communication protocols are incorporated in

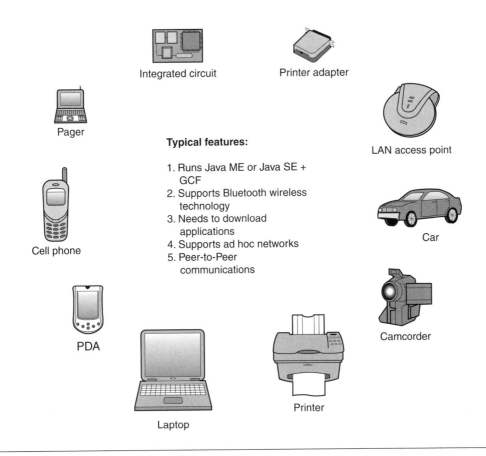

Typical features:

1. Runs Java ME or Java SE + GCF
2. Supports Bluetooth wireless technology
3. Needs to download applications
4. Supports ad hoc networks
5. Peer-to-Peer communications

Figure 2.1   Target devices for JABWT.

JABWT. Writing future Bluetooth profiles in the Java programming language enables portability across all operating systems and Bluetooth protocol stacks.

In addition to APIs for accessing the protocols, there are APIs for some Bluetooth profiles. JABWT addresses the following: GAP, SDAP, SPP, and GOEP. Detailed information on Bluetooth profiles and relations to protocols such as OBEX, RFCOMM, and L2CAP are defined in the profile's individual specification.

JABWT is based on Bluetooth specification version 1.1. However, nothing in the JABWT specification is intended to preclude operating with version 1.0–compliant stacks or hardware. In addition, if future versions are backward compatible with version 1.1, implementations of the JABWT specification also should operate on those versions of stacks or hardware.

Since JABWT was completed, two new versions of the core specification have been released. All of the features available via the JABWT are still available with these newer specifications. The one area the JABWT did not prepare for was how the profile specifications would evolve. In particular, many of the individual profile specifications utilize L2CAP PSM values previously reserved by the Bluetooth SIG. (PSM values are explained in Chapter 8.) JABWT restricts access to the reserved PSM values.

2.1.3   JABWT Use Cases

Any technology does better when more applications are created for it. Standardized APIs foster an environment to create a variety of applications. In addition, standard APIs make it possible for certain types of applications and markets that otherwise would not be possible. The portability of Java applications and standardization of JABWT facilitate the use cases discussed herein.

Peer-to-Peer Networking

Peer-to-peer networking can be defined and interpreted in many ways. For the purpose of this discussion, a peer-to-peer network is defined as a network between two or more devices whereby each device can be both a server and a client. JABWT supports peer-to-peer networking with Bluetooth wireless technology. An example of a peer-to-peer network application is a game played between two or more devices connected through Bluetooth communication.

The devices involved can belong to entirely different device classes, such as a phone and a GPS receiver using different hardware and operating systems. If these devices are JABWT enabled, the software games can be written once in the Java programming language and run on all of the devices. In addition, the device independence of these JABWT applications makes it possible to share and download these games onto different devices.

Kiosk

It is impractical for a kiosk that sells software to store different executables for the various Bluetooth devices that have been manufactured. With JABWT, an application can be written once, purchased, and executed on all Bluetooth devices that have implemented this API. This capability enables establishments such as airports, train stations, and malls to have custom applications that work best in their environment. Bluetooth devices with JABWT implemented can download these custom applications from kiosks.

Buying Soda and Bluetooth Applications through Vending Machines

Another example of the benefit of this API is a scenario in which people purchase or download Bluetooth applications to their Bluetooth devices while using the same device to purchase a soda from a vending machine. The API allows applications to be written once and run on many different Bluetooth platforms. The vending machine stores these applications and transfers them via Bluetooth transports. A game manufacturer might buy advertising space on vending machines to house a sample game. Customers purchasing soda could be given the option of downloading a free sample game, which can be upgraded later when the game is purchased.

2.2 API Characteristics and Hardware Requirements

This section describes the characteristics of JABWT and the hardware requirements followed in defining JABWT. There were two categories of hardware requirements:

- The requirements of the Java ME device
- The requirements of the Bluetooth subsystem in the device

2.2.1 JABWT Specification Characteristics

This API design was challenging because both Java technology and Bluetooth wireless technology appear in a variety of devices. It was difficult to try to cover all the devices with one API. The initial goal of the JABWT specification was to define an API that could be used by all devices that support Java ME. As stated earlier, the expert group believed that Java ME devices would be the first to implement JABWT. Hence the API was built with standard Java ME APIs and the GCF defined in CLDC. Thus JABWT can be ported to any Java platform that supports the GCF. The first two characteristics below resulted from this thinking. JSR-197 adds the GCF into Java SE platforms and will help JABWT and other Java ME APIs to be usable on other Java platforms.

The characteristics of the JABWT specification are as follows:

1. Require only CLDC libraries.

2. Scalability—ability to run on any Java platform that supports the GCF.

3. OBEX API definition independent of Bluetooth protocols. By contrast, applications written with the Bluetooth API are expected to run only on platforms that incorporate Bluetooth wireless technology. While defining the API for OBEX, the expert group recognized that OBEX could be used over a number of different transports (e.g., IrDA, USB, TCP). Therefore, the OBEX API is defined to be transport independent. The OBEX API is in a separate `javax.obex` package.

4. Use of the OBEX API without the Bluetooth API. An IrDA device could implement the `javax.obex` package and not implement the `javax.bluetooth` package, which contains the Bluetooth API.

5. Prevent applications from interfering with each other. The concept of the Bluetooth Control Center (BCC), discussed in Chapter 3, was introduced for this reason. The intent of the BCC is to allow multiple Bluetooth applications to run simultaneously and be able to access Bluetooth resources.

6. Ability of applications to be both client and server to enable peer-to-peer networking. This is one of the vital use cases for Bluetooth

wireless technology. One aspect of being a server is the ability to register services for clients to discover. Although the Bluetooth specification thoroughly addresses the client side of service discovery, the mechanisms used by server applications to register their services with a service discovery server are not standardized. The JSR-82 expert group saw the need for defining service registration in detail to standardize the registration process for the application programmer.

7. Allowance for the possibility of building Bluetooth profiles on top of the RFCOMM, L2CAP, and OBEX APIs. The expert group realized that keeping up with the growing number of Bluetooth profiles would be difficult (see Section 2.1.2).

2.2.2 Java ME Device Requirements

JABWT is not intended to be a complete solution by itself. It is an optional API based on GCF and extends a Java platform to add support for accessing Bluetooth wireless technology. As mentioned earlier, the initial target devices are CLDC based. General Java ME device requirements on which the API is designed to operate are listed below. More detailed hardware requirements for various Java ME configurations and profiles can be obtained from the respective specifications, which are available at www.jcp.org.

- 512K minimum total memory available for Java platform (ROM/Flash and RAM). Application memory requirements are additional.
- Bluetooth communication hardware, with necessary Bluetooth stack and radio. More detailed requirements are given in Section 2.2.3.
- Compliant implementation of the Java ME CLDC [15, 25] or a superset of CLDC APIs, such as the Java ME CDC [16, 26] or any flavor of Java platform with JSR-197 APIs.

2.2.3 Bluetooth System Requirements

The Bluetooth part of the JABWT implementation is not designed to access the Bluetooth hardware directly. It accesses the Bluetooth hardware through an underlying Bluetooth stack. The Bluetooth stack can be implemented in many ways, such as making it part of the JABWT implementation or writing it completely in the Java language. Typically, JABWT is to be implemented on top of a native (written in C or C++) Bluetooth stack, thus allowing native Bluetooth applications and Java Bluetooth applications to run on a system. The requirements of the underlying Bluetooth system on which this API is built are as follows:

- The underlying system is qualified in accordance with the Bluetooth Qualification Program for at least the GAP, SDAP, and SPP.
- The following layers are supported as defined in the Bluetooth specification version 1.1, and the JABWT has access to them. (Support has also been shown in the latest Bluetooth core specification.)

   SDP

   RFCOMM

   L2CAP

- The BCC is provided by either the Bluetooth stack or system software. The BCC is a "control panel"–like application that allows a user or an original equipment manufacturer (OEM) to define specific values for certain configuration parameters in a stack. The details of the BCC are discussed in Chapter 3.

Unlike the Bluetooth part of the API, the OBEX API can either be implemented completely in the Java programming language within the JABWT implementation or can use the OBEX implementation in the underlying Bluetooth stack. If OBEX is being implemented on another transport, the OBEX API can use the OBEX implementation over that transport system.

2.3 Scope

The Bluetooth specification covers many layers and profiles, and it is not possible to include all of them in this API. Rather than try to address all of them, the JABWT expert group agreed to prioritize the API functions on the basis of size requirements and the breadth of usage of the API. Moreover, under the JCP rules, when JABWT is implemented, all portions of the API must be implemented (i.e., if the `javax.bluetooth` package is implemented, then RFCOMM, SDP, and L2CAP must be implemented; if `javax.obex` is implemented, then OBEX must be implemented). The Bluetooth specification is different because it is flexible about the parts of the Bluetooth specification a device manufacturer chooses to implement. The expert group addressed areas considered essential to achieving broad usage and areas expected to benefit from having support in the Java language. As stated earlier, these APIs are aimed at small, resource-constraint devices of different classes. The Headset Profile [47] or the Dial-Up Networking Profile [32] will likely be developed by a device manufacturer as an application native to the system software. For the first version of JABWT, support for voice channels and telephony control–related areas were not included in JABWT. The basic Bluetooth profiles and fundamental protocol layers required to help build future profiles were included. In addition, service registration was defined in detail.

Figure 2.2 shows that JABWT applications have access to some but not all of the functionality of the Bluetooth protocol stack. The bottom of Figure 2.2 reproduces Figure 1.4 from Chapter 1, which shows the layers in a Bluetooth stack. In Figure 2.2, interface points have been added to represent the capabilities or functions of protocols that could potentially be used by applications. In Figure 2.2 dashed arrows connect the JABWT application at the top of the figure with interface points on the protocols in the Bluetooth protocol stack. An arrow connecting to an interface point indicates that JABWT applications have access to the functionality represented by that interface point. As shown in Figure 2.2, JABWT provides access to capabilities of the following Bluetooth protocols:

- L2CAP
- RFCOMM
- SDP
- OBEX
- LMP

JABWT does not provide APIs for the following Bluetooth protocols:

- Audio (voice) transmissions over voice channels
- TCS Binary
- BNEP

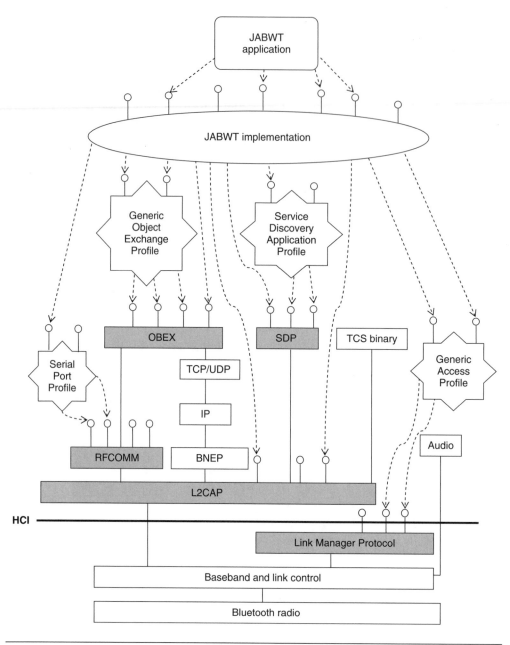

Figure 2.2   JABWT provides access to selected functionality of the Bluetooth stack.

Even when it does provide access to a Bluetooth protocol layer, JABWT might not provide access to all of the functions provided by that layer. For example, JABWT applications have access to connection-oriented L2CAP channels but do not have access to connectionless L2CAP channels. This possibility is indicated in Figure 2.2 by an interface point that has nothing connected to it.

Figure 2.2 shows that in addition to providing access to the functionality of Bluetooth protocols, JABWT provides access to the functionality specified by the Bluetooth profiles. The star shapes in Figure 2.2 represent Bluetooth profiles. JABWT applications have access to selected functionality from the following Bluetooth profiles:

- GAP
- SDAP
- SPP
- GOEP

In functional terms, JABWT provides the following Bluetooth capabilities:

- Registering services
- Discovering devices and services
- Establishing RFCOMM, L2CAP, and OBEX connections
- Conducting the above three activities in a secure manner

The following capabilities were considered to be outside the scope of JABWT. However, there is no incompatibility between JABWT and these functions, so JABWT applications may have access to these functions on some devices:

- Layer management: Many aspects of layer management are system specific and are difficult to standardize, such as power modes and park mode.
- Downloading and storing applications: These features are implementation specific and therefore are not defined in JABWT. Over-the-air provisioning is being addressed in other JSRs.

2.4 Summary

This chapter discusses the goals, capabilities, characteristics, and scope of JABWT. Although the Bluetooth specification defines a standard for over-the-air communication, JABWT standardizes software APIs for use by Bluetooth applications. One of the design goals for this API was to make it possible to write Bluetooth profiles in the Java programming language using JABWT. For this reason, JABWT provides support for the most basic Bluetooth protocols and the most basic Bluetooth profiles.

The following are some of the key characteristics of JABWT:

- It uses the CLDC generic connection framework.
- It requires a BCC for system control.
- It provides a definition for service registration.
- It defines an OBEX API that is transport independent.

JABWT defines two separate Java packages, `javax.bluetooth` and `javax.obex`. Under JCP licensing rules, these JABWT packages must be implemented exactly as defined without addition or removal of public classes, interfaces, or methods. The underlying Bluetooth system needs to be qualified for GAP, SDAP, and SPP. In addition, the underlying Bluetooth system must provide access to SDP, RFCOMM, and L2CAP. Section 2.3 discusses the scope of the JABWT specification. The three main areas that JABWT does not currently support are audio over SCO links, TCS-BIN, and BNEP.

JABWT is aimed mainly at Java ME devices. In conjunction with JSR-197, which adds optional support for the GCF to Java SE, JABWT also is well suited for Java SE devices.

3 High-Level Architecture

CHAPTER

This chapter discusses the high-level architecture of JABWT. The chapter introduces the following:

- Architecture of the JABWT specification
- The Bluetooth Control Center
- A simple JABWT application

3.1 Architecture of JABWT

The functionality provided by JABWT falls into three major categories:

1. Discovery
2. Communication
3. Device management

Discovery includes device discovery, service discovery, and service registration. Communication includes establishing connections between devices and using those connections for Bluetooth communication between applications. These connections can be made over several different protocols, namely RFCOMM, L2CAP, and OBEX. Device management allows for managing and controlling these connections. It deals with managing local and remote device states and properties. It also facilitates the security aspects of connections. JABWT is organized into these three functional categories.

3.1.1   CLDC, MIDP, and JABWT

JABWT depends only on the CLDC and uses the GCF. But CLDC does not necessarily make a complete solution. It is usually coupled with a Java ME profile such as the MIDP [18, 27]. MIDP devices are expected to be the first class of devices to incorporate JABWT.

Figure 3.1 is an example of how the APIs defined in JABWT fit in a CLDC + MIDP architecture. Although shown here on an MIDP device, JABWT does not depend on MIDP APIs. The lowest-level block in the figure is the system software or host operating system. The host operating system contains the host part of the Bluetooth protocol stack and other libraries used internally and by native applications of the system. Native Bluetooth applications interface with the operating system directly, as shown in Figure 3.1. The CLDC/KVM implementation sits on top of the host system software. This block provides the underlying Java execution environment on which the higher-level Java APIs can be built. The figure shows two such APIs that can be built on top of CLDC:

- JABWT, the set of APIs specified by JSR-82
- MIDP, the set of APIs defined by JSR-37 and JSR-118

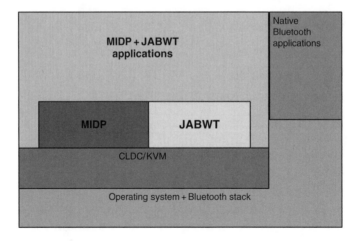

Figure 3.1   CLDC + MIDP + Bluetooth architecture diagram.

As shown in Figure 3.1, an application written for an MIDP + JABWT device can access MIDP, JABWT, and CLDC layers directly.

These diagrams describe the architecture of the JABWT reference implementation developed by us and our team at Motorola. Other JABWT implementations may involve different components or have their components layered in a different way from that shown.

3.1.2 Java Packages

As stated in Chapter 2, JABWT essentially defines two separate APIs. Hence two Java packages are defined. The packages are as follows:

1. javax.bluetooth
2. javax.obex

The OBEX API is defined independently of the Bluetooth transport layer and is packaged separately. Each of the two Java packages represents a separate optional package, the implication being that a CLDC implementation can include neither of them, one of them, or both of them. The `javax.bluetooth` package contains the Bluetooth API, and the `javax.obex` package contains the APIs for OBEX.

Figure 3.2 shows the package structure. The `javax.obex` and `javax.bluetooth` packages depend on the `javax.microedition.io` package, which contains the GCF.

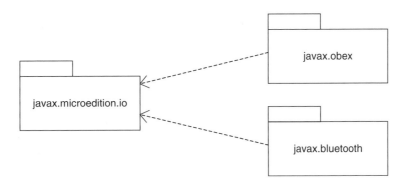

Figure 3.2 Package structure.

3.1.3   Client and Server Model

An overview of the Bluetooth client and server model is given in this section. Additional details are provided in later chapters.

A Bluetooth service is an application that acts as a server and provides assistance to client devices via Bluetooth communication. This assistance typically takes the form of a capability or a function unavailable locally on the client device. A printing service is one example of a Bluetooth server application. Additional examples of Bluetooth server applications can be found at the Bluetooth SIG Web site (www.bluetooth.com and www.bluetooth.org): LAN access servers, file and object servers, synchronization services, and so on. JABWT developers can create Bluetooth server applications to implement one of the Bluetooth profiles or to implement their own custom service. These services are made available to remote clients by the definition of a service record that describes the service and the addition of that service record to the service discovery database (SDDB) of the local device.

Figure 3.3 illustrates the Bluetooth components involved in service registration and service discovery. The SDP is a Bluetooth protocol for discovering the services provided by a Bluetooth device. A server application adds a service record to the SDDB. The Bluetooth stack provides an SDP server, which maintains this database of service records. Service discovery clients use SDP to query the SDP server for any service records of interest. A service record provides sufficient information to allow an SDP client to connect to the Bluetooth service on the server device.

After registering a service record in the SDDB, the server application waits for a client application to initiate contact with the server to access the service. The client application and the server application then establish a Bluetooth connection to conduct their business.

Although the Bluetooth specification was used as a guide for defining the capabilities that should be offered in JABWT, defining the capabilities of the server applications was more difficult, because the Bluetooth specifications do not specify:

- How or when server applications register service records in the SDDB
- What internal format or database mechanism is used by the SDDB
- How server applications interact with the Bluetooth stack to form connections with remote clients

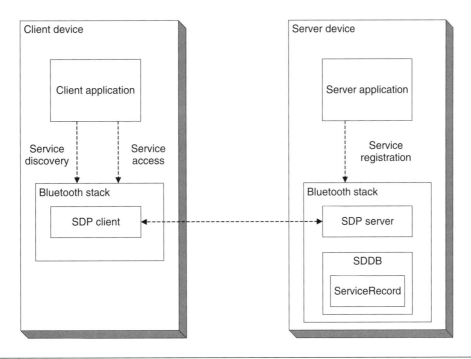

Figure 3.3 Bluetooth components involved in service discovery.

These aspects of server applications are outside the scope of the Bluetooth specification and are likely to vary from one Bluetooth stack implementation to another. They do not require standardization to ensure interoperability of Bluetooth devices from different manufacturers. However, the JABWT specification for service registration allows server applications to take full advantage of Bluetooth communications. Standardization of server registration is an additional benefit JABWT brings to the programming community.

JABWT defines the following division of responsibilities among the server application, the client application, and the Bluetooth stack.

Typical responsibilities of a Bluetooth server application are to:

- Create a service record describing the service offered by the application
- Add a service record to the server's SDDB to make potential clients aware of this service

- Register the Bluetooth security measures associated with this service that should be enforced for connections with clients
- Accept connections from clients that request the service offered by the application
- Update the service record in the server's SDDB if characteristics of the service change
- Remove or disable the service record in the server's SDDB when the service is no longer available

Typical responsibilities of a Bluetooth client application are to:

- Use SDP to query a remote SDDB for desired services
- Register the Bluetooth security measures associated with this service that should be enforced for connections with servers
- Initiate connections to servers offering desired services

The Bluetooth stack is assumed to provide the following capabilities for Bluetooth server applications:

- A repository for service records that allows servers to add, update, and remove their own service records
- Connections with remote client applications

The Bluetooth stack is assumed to provide the following capabilities for service discovery clients:

- Search and retrieval of service records stored in the server's SDDB (i.e., acting as an SDP server)
- Connections to server applications

Peer-to-Peer Applications

Although it is important to understand the distinction between a Bluetooth client application and a Bluetooth server application, it is possible for the same Bluetooth application to play both the client role and the server role. It is one of the stated goals of JABWT to support peer-to-peer applications in which the peer-to-peer application is capable of being both server and client. For example, it is not likely that a two-person

Bluetooth game would be sold in client and server versions. Instead, the game software would do both of the following:

- Initiate attempts to connect to nearby devices that have the same game (client)
- Accept connections requested by nearby devices with the same game (server)

Whereas JABWT tends to describe the client and server techniques separately, these techniques are not incompatible, and applications can use them both. Service discovery and service registration are discussed in more detail in Chapter 7.

3.1.4 Device Properties

Various Bluetooth products need to be configured differently depending on the product and the market. A set of device properties facilitates such variations and differentiations. JABWT defines system properties that may be retrieved by a call to `LocalDevice.getProperty()`. These properties do either of the following:

- Provide additional information about the Bluetooth system; that is, the capabilities of the device or the underlying Bluetooth protocol stack.
- Define restrictions placed on an application by an implementation. The device manufacturer may want to restrict certain capabilities for various reasons.

An example of these device properties is `bluetooth.connected.devices.max`, which indicates the maximum number of Bluetooth devices that can connect to this device. Device properties are discussed in Chapter 6.

3.2 Bluetooth Control Center

The BCC is part of the JABWT specification, but it does not have any Java APIs that provide direct access to it. In other words, the BCC is a concept defined by the JABWT specification, which is part of a JABWT implementation. The need for the BCC arises from the desire to prevent one application from adversely affecting another application. The BCC is the central authority for local Bluetooth device settings. The details of the BCC are left to the implementation. It may be an interactive application with a user interface or an application that provides no user interaction. The BCC may be a native application, an application with a separate private Java API, or simply a group of settings specified by the manufacturer.

The BCC performs three specific tasks:

1. Resolves conflicting requests between applications
2. Enables modifications to the properties of the local Bluetooth device
3. Handles security operations that may require user interaction

Each of these tasks is discussed individually in the next sections.

As Figure 3.4 shows, the BCC is not directly accessible with JABWT applications. Instead, the JABWT implementation issues requests through the BCC to the Bluetooth stack. The BCC also can be used by native applications. The BCC can prevent conflicting requests between the JABWT applications and the native applications.

How does a user modify the values of the BCC? This procedure also is up to an implementation of the BCC. It is expected that most implementations will use a native application to manipulate the settings in the BCC.

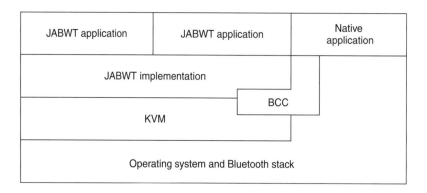

Figure 3.4   How the BCC fits into a JABWT implementation.

3.2.1   Conflict Resolution

The JABWT specification allows a great deal of flexibility within a JABWT implementation. This feature was included for two reasons. First, the flexibility resulted from the desire to allow the JABWT implementation to be ported to a large number of Bluetooth stacks and radios. Second, JABWT implementations are able to differentiate themselves on the basis of the policies the implementation enforces.

Because multiple applications are able to run and access the local Bluetooth device at the same time, conflicting requests can be made to the Bluetooth device. As far as JABWT is concerned, two types of requests can conflict with another application using the same Bluetooth device. First, two applications may request different security settings on a link. (Bluetooth security is described in Section 4.1.) Second, two applications may request to set the device into two different discoverable modes. (Discoverable modes are described in Chapter 6.) The BCC is responsible for resolving these conflicting requests.

3.2.2   Modifying Device Properties

Although JABWT allows an application to retrieve certain properties of the local device, no methods within JABWT allow direct modification of a device's properties. In particular, setting of the friendly name, the class of device record, the list of pre-known devices, the list of trusted devices, the minimum security requirements, and support for the different connectable/discoverable modes are handled by the BCC. (Each of these concepts is described later in this book.) The friendly name is a user-friendly name given to a Bluetooth device. The name does not uniquely identify a Bluetooth device but provides a name of a device that can be displayed to a user instead of a device's Bluetooth address. For example, a user named Bob may assign his phone the friendly name "Bob's Cell."

3.2.3   User Interaction

Certain operations within Bluetooth security may require input from the user of a device outside the scope of the application. The BCC is responsible for retrieving this information from the user and injecting the information into the Bluetooth security process. What type of information can the BCC retrieve from the user? It can range from a PIN to simply responding to a permission request.

3.2.4 BCC on Devices with No User Interface

Because JABWT is based on CLDC, there is no guarantee that a user interface (UI) is available on the device. In this situation, the OEM or device manufacturer is expected to set the BCC configuration in the device. Actions that require user interaction are more complicated. A BCC on a non–graphical user interface (GUI) device might not support these types of actions or can specify the responses to these actions when the device is manufactured.

3.3 Simple JABWT Application

Before describing the details of the classes and methods defined in JABWT, the traditional "Hello, World" application is shown. This example shows how code is presented in the remainder of the book. Because Bluetooth technology is a wireless radio technology, developing applications requires hardware or a simulator. To enable readers to try out the code in this book, the following section also describes how to set up a development environment that makes it possible to build and test JABWT code in a simulated environment.

3.3.1 Development Tools

Developing and testing Java ME applications, especially testing on a device, can be a complicated process. Device testing is complicated due to a general lack of debug tools and the effort it takes to download and install an application. Therefore device simulators have been developed to allow developers to create and debug applications on a desktop computer before testing them on a device. A common tool for Java ME development is the Sun Java Wireless Toolkit available at java.sun.com/javame. The Wireless Toolkit provides software emulation of devices that support a variety of Java ME specifications. The Wireless Toolkit also provides support for building, packaging, and testing Java MIDlets. (Support for JSR-82 is now available in the Wireless Toolkit, but will not be used to test or demo the code in this book.)

Another tool used in this book is the Motorola Java ME SDK for Motorola OS Products. The Motorola Java ME SDK for Motorola OS Products is available at www.motodev.com. (This book utilized version 6.4 of the Java ME SDK for Motorola OS Products to test the code examples.) The SDK provides support for device emulation and emulation of the Bluetooth stack via Rococo Software's Impronto™ Simulator.

Follow the instructions to install the Sun Java Wireless Toolkit, Java ME SDK for Motorola OS Products, and the Impronto Simulator. The instructions to install the Impronto Simulator are available via a ReadMe.txt in the Impronto_Simulator folder in the Motorola SDK installation directory.

3.3.2   Sample Application

Before introducing the details of JABWT, let's take a look at how simple it is to get up and running with JABWT. A simple "Hello, World" application follows. The `HelloClient` MIDlet locates a `HelloServer` MIDlet and sends the text "Hello, World" to the server to be displayed by the `HelloServer` on its screen.

To start the project, start the Sun Java Wireless Toolkit and create a new project. Provide a project name and the name of the MIDlet class. (See Table 3.1 for the project name and MIDlet class name for the sample application.) After providing this information, make sure JSR-82 is selected in the API Selection tab and press the ok button. After selecting ok, a new project is created by the Wireless Toolkit. The output window in the Wireless Toolkit will display where the project source directory is located (see Figure 3.5).

Table 3.1   Project Name and MIDlet Class Name for the Sample Application

Project Name	MIDlet Class Name
HelloServer	com.jabwt.book.HelloServer
HelloClient	com.jabwt.book.HelloClient

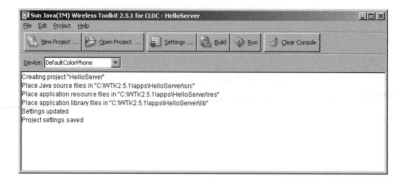

Figure 3.5   The Sun Java Wireless Toolkit after creating the HelloServer project.

Before showing the JABWT code, the `BluetoothMIDlet` class is introduced. `HelloClient` and `HelloServer` use this class as a building block. `BluetoothMIDlet` starts a processing thread and destroys the MIDlet when a `Command` is selected. (The following class must be placed in each project source directory in this book since it is reused by each chapter in this book.)

```
package com.jabwt.book;

import java.lang.*;
import java.io.*;
import javax.microedition.io.*;
import javax.microedition.lcdui.*;
import javax.microedition.midlet.*;
import javax.bluetooth.*;

public class BluetoothMIDlet extends MIDlet implements
  Runnable, CommandListener {
  public BluetoothMIDlet() {}
  /**
   * Starts a background thread when the MIDlet is
   * started.
   */
  public void startApp()
    throws MIDletStateChangeException {
    new Thread(this).start();
```

```
    }
    public void pauseApp() {}
    public void destroyApp(boolean unconditional) {}
    public void run() {}
    /**
     * Destroys the MIDlet when a Command occurs.
     */
    public void commandAction(Command c, Displayable d) {
      notifyDestroyed();
    }
}
```

The next step is to write the `HelloServer` code. The `run()` method of `HelloServer` does all the work. It makes the server device discoverable so that the client can find the server. Next, the `run()` method waits for a client to connect and reads all the data sent from the client. The `run()` method displays the data sent from the client on the screen.

```
package com.jabwt.book;

import java.lang.*;
import java.io.*;
import javax.microedition.lcdui.*;
import javax.microedition.io.*;
import javax.bluetooth.*;

public class HelloServer extends BluetoothMIDlet {
  /**
   * Creates a server object. Accepts a single
   * connection from a client and prints the data
   * sent from the client to the screen.
   */
  public void run() {
    // Create a Form and add the Exit command to the Form
    Form f = new Form("Server");
    f.addCommand(new Command("Exit", Command.EXIT, 1));
    f.setCommandListener(this);
    Display.getDisplay(this).setCurrent(f);
```

```
try {
  // Make the local device discoverable for the
  // client to locate
  LocalDevice local = LocalDevice.getLocalDevice();
  if (!local.setDiscoverable(DiscoveryAgent.GIAC)) {
    f.append("Failed to change to the " +
      "discoverable mode");
    return;
  }
  // Create a server connection object to accept
  // a connection from a client
  StreamConnectionNotifier notifier =
    (StreamConnectionNotifier)
    Connector.open("btspp://localhost:" +
      "86b4d249fb8844d6a756ec265dd1f6a3");
  // Accept a connection from the client
  StreamConnection conn = notifier.acceptAndOpen();
  // Open the input to read data from
  InputStream in = conn.openInputStream();
  ByteArrayOutputStream out = new
    ByteArrayOutputStream();
  // Read the data sent from the client until
  // the end of stream
  int data;
  while ((data = in.read()) != -1) {
    out.write(data);
  }
  // Add the text sent from the client to the Form
  f.append(out.toString());
  // Close all open resources
  in.close();
  conn.close();
  notifier.close();
} catch (BluetoothStateException e) {
  f.append("BluetoothStateException: ");
```

```
        f.append(e.getMessage());
      } catch (IOException e) {
        f.append("IOException: ");
        f.append(e.getMessage());
      }
    }
  }
}
```

To verify the code was properly copied from the book, build the code using the Wireless Toolkit by pressing the "Build" button. Once the build succeeds, package the build using the "Project->Package-> Create Package" menu option.

After the `HelloServer` MIDlet is created, the `HelloClient` MIDlet must be written to send the "Hello, World" message to the server. All the work for the `HelloClient` MIDlet occurs in the `run()` method. The `run()` method uses the `selectServices()` method to discover the `HelloServer`. After discovering the server, the `HelloClient` connects to the server and sends the text. Figure 3.8 shows a successful run of the `HelloClient` and `HelloServer` MIDlets.

```
package com.jabwt.book;

import java.lang.*;
import java.io.*;
import javax.microedition.io.*;
import javax.microedition.lcdui.*;
import javax.bluetooth.*;

public class HelloClient extends BluetoothMIDlet {
  /**
   * Connects to the server and sends 'Hello, World'
   * to the server.
   */
  public void run() {
    // Creates the Form and adds the Exit Command to it
    Form f = new Form("Client");
    f.addCommand(new Command("Exit", Command.EXIT, 1));
    f.setCommandListener(this);
    Display.getDisplay(this).setCurrent(f);
```

```
try {
  // Retrieve the connection string to connect to
  // the server
  LocalDevice local =
    LocalDevice.getLocalDevice();
  DiscoveryAgent agent =local.getDiscoveryAgent();
  String connString = agent.selectService(new
    UUID("86b4d249fb8844d6a756ec265dd1f6a3", false),
    ServiceRecord.NOAUTHENTICATE_NOENCRYPT, false);
  if (connString != null) {
    try {
      // Connect to the server and send 'Hello, World'
      StreamConnection conn = (StreamConnection)
        Connector.open(connString);
      OutputStream out = conn.openOutputStream();
      out.write("Hello, World".getBytes());
      out.flush();
      out.close();
      conn.close();
      f.append("Message sent correctly");
    } catch (IOException e) {
      f.append("IOException: ");
      f.append(e.getMessage());
    }
  } else {
    // Unable to locate a service so just print an error
    // message on the screen
    f.append("Unable to locate service");
  }
} catch (BluetoothStateException e) {
  f.append("BluetoothStateException: ");
  f.append(e.getMessage());
}
  }
}
```

After building and packaging the `HelloClient` MIDlet, the next step is to configure two Bluetooth devices in the simulator. (Remember to copy the `BluetoothMIDlet` class to the HelloClient project's source directory.) Start the simulator according to the instructions in the Motorola SDK. After starting the simulator, create two new Bluetooth devices named "client" and "server" (see Figure 3.6).

Once the two devices are configured in the simulator, the Motorola LaunchPad application can be started. Instead of starting the application from the Start Menu, open two MS-DOS command prompts. Within each command prompt, change to the "C:\Program Files\Motorola\ Motorola Java ME SDK v6.4 for Motorola OS Products" or the directory in which Motorola LaunchPad was installed. In one MS-DOS prompt, enter the command "set SIM_FRIENDLY_NAME=server" without the quotes. In the other MS-DOS prompt, enter the command "set

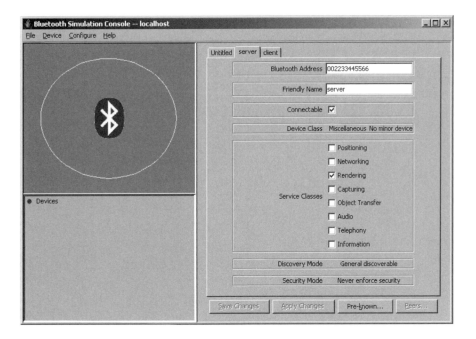

Figure 3.6   Configure the server and client devices in the Impronto Simulator.

SIM_FRIENDLY_NAME=client" without the quotes. (The SIM_FRIENDLY_
NAME identifies a unique device in the Impronto Simulator. Completing
these two steps sets one LaunchPad session as the client and the other as
the server.)

After properly setting up the environment, you can start Launch-
Pad by invoking the Launchpad.exe executable (see Figure 3.7). To run
the client and server programs, select a Motorola handset that supports
JSR-82 such as the MOTORIZR Z3 and enter the full path to the
`HelloClient.jad` file or `HelloServer.jad` file that was created by
the Wireless Toolkit. The client and server can then be launched by
pressing the launch button. Figure 3.8 shows the two applications
running.

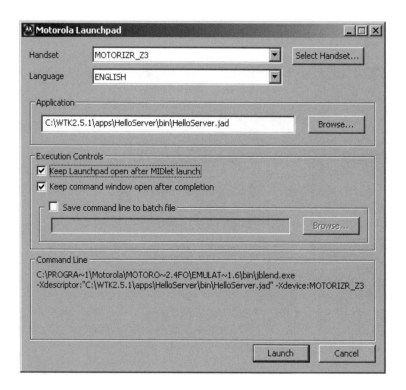

Figure 3.7   Running the `HelloServer` MIDlet from the Motorola Launchpad.

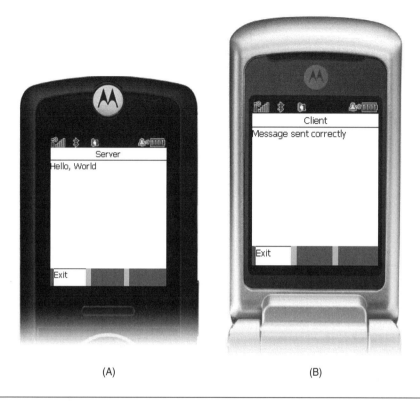

Figure 3.8 A run using the Impronto Simulator. (A) `HelloServer`; (B) `HelloClient` (emulation only).

3.4 Summary

This chapter presents the high-level architecture of JABWT to set the stage for the detailed API discussions in the coming chapters. Because JABWT is expected to be implemented first on CLDC/MIDP devices, Section 3.1.1 describes how JABWT can fit into a CLDC/MIDP device. A client-server model is basic to the operation of Bluetooth wireless technology, and that client-server model is reflected in JABWT. JABWT goes a step further than the Bluetooth specification in standardizing service registration. To allow for variations in Bluetooth product configuration, JABWT defines configurable system properties.

JABWT introduces the concept of a BCC to allow for system control and monitoring. Some form of BCC must be part of all JABWT implementations. However, the details of the BCC are left to the JABWT implementation. The three main tasks the BCC performs are conflict resolution, modification of system properties, and user interaction.

Section 3.3 presents a simple "Hello, World" JABWT application to introduce the APIs discussed in the following chapters.

4 RFCOMM

This chapter covers the following topics:

- What is the SPP?
- Why use RFCOMM?
- How do you establish an RFCOMM connection?
- How do you create a new RFCOMM service?
- Communicating over RFCOMM
- Bluetooth security in RFCOMM
- Specifying the master and slave device

4.1 Overview

The SPP is the Bluetooth profile that realizes an RFCOMM connection between two devices. The SPP is defined as a building block profile. This means that other Bluetooth profiles are built on the SPP. Figure 4.1 shows some of the Bluetooth profiles built on the SPP. The Generic Object Exchange Profile (GOEP) is shown inside the Serial Port Profile box in Figure 4.1, which indicates that the GOEP is built on top of the SPP. In basic terms, the SPP profile defines how two Bluetooth devices establish two-way, reliable communication with the RFCOMM protocol.

The RFCOMM protocol is an emulation of an RS-232 serial port connection between two devices over a wireless link. Within JABWT, communicating with a remote device using RFCOMM is similar to communicating over a socket connection. In other words, data is sent between devices via streams. In most situations, RFCOMM should be the protocol to use within a JABWT application. This is because serial communication is widely used and the API is simple to use.

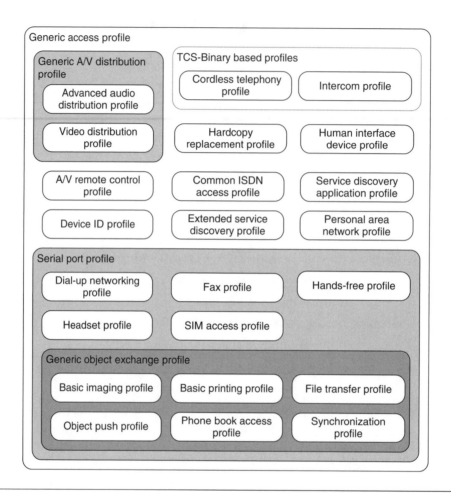

Figure 4.1   Bluetooth profiles defined by the Bluetooth SIG [42, 43, 44, 45].

Before continuing, it is important to understand some of the termi-
nology used within Bluetooth networking. Even though Bluetooth networking is a wireless technology, only a single "physical" link exists
between any two Bluetooth devices. Although there may be only a single
link, there may be multiple connections between the two devices over
this link (Figure 4.2). The situation is similar to the wired networking
world. Although there is only a single Ethernet cable between two
devices, there may be multiple connections between the two devices.

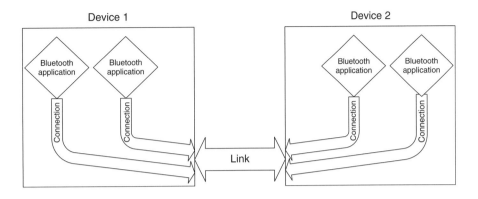

Figure 4.2   Multiple Bluetooth connections can exist over a single Bluetooth link.

Bluetooth wireless technology provides different levels of security over a Bluetooth link. There are four types of Bluetooth security: pairing, authentication, encryption, and authorization. Pairing is the first step in the process of Bluetooth security. When two devices come into contact with one another for the first time and want to use security, the devices must establish a shared secret used for authentication and encryption. Pairing requires the user of each device to input a common code or PIN into each device. The PIN is then used to do an initial authentication of both devices. After the initial pairing, a shared secret is established and is stored within the Bluetooth device to allow authentication of both devices in the future without the need for the pairing process. Figure 4.3 shows how two devices can retrieve the PIN to complete the pairing process. The pairing process is transparent to the application. It is the responsibility of the BCC to retrieve the PIN from the user or determine what the PIN should be.

Bluetooth authentication verifies the identity of one device to another device using a challenge and response scheme. Bluetooth authentication does not authenticate users but authenticates devices. When device A wants to authenticate device B, device A sends a challenge to device B (Figure 4.4). When it receives this challenge, device B applies the shared secret to the challenge and sends the result to device A. Device A then combines the challenge that was sent with its shared secret and compares the result with the result sent from device B. Although it

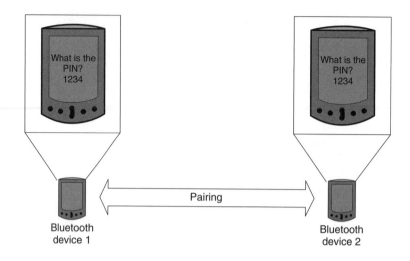

Figure 4.3   For two devices to complete the pairing process, a common PIN must be entered.

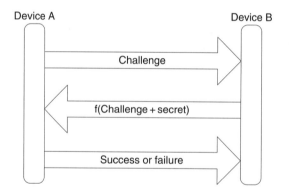

Figure 4.4   Device A attempts to authenticate device B.

authenticates device B to device A, this process does not authenticate device A to device B. The same process must be used to authenticate device A to device B. To perform authentication, device A and device B must complete the pairing process so that the shared secret can be established.

Once the authentication process has been completed, encryption can be turned on. Figure 4.5 shows an example of what it means for a

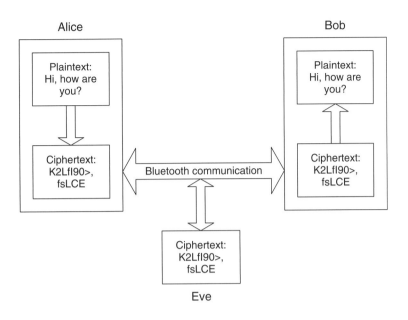

Figure 4.5   Example of encryption.

link to be encrypted. Encryption is used to prevent an eavesdropper, Eve, from intercepting communication between two entities, Alice and Bob. When one device wants to turn on encryption, it must ask the other Bluetooth device to do so also. If the other device accepts the request, all packets between the devices are encrypted. If the other device rejects the request, the connection is closed. Unlike the mechanism of authentication, it is not possible for communications sent from device A to device B to be encrypted while communications sent from device B to device A are unencrypted.

Another option within Bluetooth security is authorization. Authorization is the process of determining whether a connection request from a specific Bluetooth device should be granted. Authorization is completed on a connection-by-connection basis. The Bluetooth specification has also defined the concept of a trusted device. What is a trusted device? A trusted device is a device that is automatically granted authorization when authorization is requested. In other words, a trusted device is authorized to connect to any service on the local device. When a trusted device connects to a service that requires authorization, the request is automatically accepted without the BCC asking the user if the device is authorized to use the service. The BCC is in charge of maintaining the list of trusted devices. When an authorization request is received by the BCC for a nontrusted device, the BCC requests the user to grant or deny the connection.

Each level of security builds on the previous level. Authentication requires pairing. Encryption and authorization require authentication. JABWT enforces these requirements. If encryption is requested on a link and the link has not been authenticated, the JABWT implementation authenticates the remote device before encrypting the link.

4.2 API Capabilities

No new methods or classes were defined for RFCOMM communication; instead, existing classes and interfaces from the GCF were used. As with all Java ME communication, using RFCOMM starts with the GCF (see Figure 4.6). A well-defined connection string is passed to `Connector.open()` to establish the connection. For client connections, a `StreamConnection` object is returned from `Connector.open()`. `Connector.open()` returns a `StreamConnectionNotifier` object if a server connection string is used. Once a connection has been established between a client and a server, the client and server communicate via `InputStreams` and `OutputStreams`.

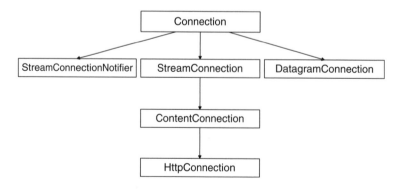

Figure 4.6   GCF defined by CLDC.

JABWT allows security to be modified by an application at two different times. Security can be modified when a connection is first established and after the connection is established. To set security when a connection is established, three parameters can be added to the connection string passed to `Connector.open()`. (Section 6.3.3 describes how to change security on a connection after the connection is established.) The BCC is responsible for verifying that these parameters are acceptable and resolving conflicting security requests. In other words, all security requests on a link must go through the BCC.

Resolving conflicting security requests is a complicated problem because changing security in an unexpected way can cause serious problems for an application. For example, a banking application may allow a user to pay for groceries over a Bluetooth link. The application transmits the user's bank account number over an encrypted Bluetooth link. If the link is not encrypted, someone listening on the Bluetooth link could steal the user's bank account number.

Although JABWT does not specify how conflicting security requests should be handled, it is expected that most implementations prevent one application from decreasing the security on a link as long as another application believes the link has a certain security level. This expectation is based on the fact that an implementation that does not enforce this policy would leave an application with no expectations of security at any time. This expectation leads to three possible implementations. First, the BCC enforces the same level of security on all applications. If an application requests a different level of security, the application's request fails. Second, the first application to request security on a link receives its requested level of security. If a second application comes along and requests a higher level of security, the second application's request fails. The third approach is the most complicated. As in the second approach, the first application receives the level of security it requests on a link. If the second application requests a higher level of security, the JABWT implementation attempts to increase the level of security on the link. If the request succeeds, the second application receives its connection. If the second application requests a lower level of security, the second application receives a connection with the first connection's higher level of security.

Within every Bluetooth link between two devices, one of the devices is considered the master and the other the slave of the

Device A   Device B
(Master)   (Slave)

Figure 4.7   Master connects to slave.

connection. The master device drives the frequency-hopping sequence used by both devices during the wireless connection. (The frequency hopping is done for security reasons and to minimize interference with other wireless devices.) For most applications, the master and slave configuration is not important, but if a developer is implementing a Bluetooth profile, the developer may need to consider which device is the master and which is the slave. Another reason a developer would like to configure a device to be master is to enable the device to form a *piconet*. A piconet is a network of up to seven Bluetooth devices. Being the master allows a device to establish additional connections to other devices in the area. The device that initiates a connection starts out as the master of the connection. The device with the service being connected to it is initially the slave (Figure 4.7).

4.3   Programming with the API

All RFCOMM communication begins with `Connector.open()` and a valid connection string. All connection strings passed to `Connector.open()` are of the form

    {scheme}:{target}{params}

To use RFCOMM, the {scheme} used for both client and server connection strings is `btspp`. The {target} and {params} are different depending on whether the connection is a client or a server.

    In addition to the {scheme} being the same for client and server connections, there are similar {params} for both types of connections. Table 4.1 lists all the valid {params} that may be used in an RFCOMM, L2CAP, and OBEX over RFCOMM connection string along with the valid values for each of the {params}. All other values would cause an

Table 4.1   Valid Parameters for RFCOMM Connection Strings

Name	Description	Valid Values	Client or Server
master	Specifies whether this device must be the master of the connection	true, false	Both
authenticate	Specifies whether the remote device must be authenticated before establishing a connection	true, false	Both
encrypt	Specifies whether the link must be encrypted	true, false	Both
authorize	Specifies whether all connections to this device must receive authorization to use the service	true, false	Server
name	Specifies the ServiceName attribute in the service record (service records are explained further in Chapter 7)	Any valid string	Server

`IllegalArgumentException` to be thrown by `Connector.open()`. Each of these {params} is optional and therefore does not have to be included in the connection string.

The parameters that set the security requirements of RFCOMM, L2CAP, and OBEX over RFCOMM are `authenticate`, `encrypt`, and `authorize`. These parameters have the value `true` or `false`. The security parameters do not have to be set. If the parameter is not included in the string, the implementation interprets the parameter as `false` unless another parameter requires this parameter to be `true`. For example, if `encrypt` is set to `true` and `authenticate` is not part of the connection string, the link is authenticated even though it was not set to `true` in the connection string, because encryption requires authentication.

Certain combinations of parameters are not valid. `Authenticate` cannot be set to `false` if `encrypt` or `authorize` is set to `true`. If an invalid combination of parameters is passed to `Connector.open()`, a `BluetoothConnectionException` is thrown. If the authentication, encryption, or authorization request fails during the establishment of the connection, a `BluetoothConnectionException` also is thrown.

To enable implementing profiles over a JABWT implementation, JABWT provides a way for a service to request that the local device be the master of the connection. When the service makes the request to `Connector.open()` to retrieve the service's notifier object, the connection string to produce the notifier object takes another parameter. The `master` parameter has two valid values: `true` and `false`. If the `master` parameter is set to `true`, then to use the service, the device initiating the connection must give up the master role. If the `master` parameter is `false`, the device does not care whether it is the master or the slave. There is no API to force a device to be the slave. The `master` parameter is valid for client and server connection strings. Not all devices support changing the master of a connection. If the device does not support changing the master of a connection, then a `BluetoothConnection-Exception` is thrown. (For more information on connection strings, see Chapter 8 for L2CAP and Chapter 5 for OBEX.)

The `name` parameter is a server-specific parameter. The `name` parameter specifies the ServiceName attribute in the service record. The `name` parameter can have a value of any valid string.

4.3.1   Establishing a Server Connection

To establish a server connection, a valid server connection string must be passed to `Connector.open()`. The {scheme} to use is `btspp`. The {target} for server connections is the keyword `//localhost` followed by a colon and the universally unique identifier (UUID) for the service to add to the service record. Not only is a `StreamConnectionNotifier` created by `Connector.open()`, but also a basic service record is created. It is not registered into the service record database until `acceptAndOpen()` is called on the `StreamConnectionNotifier` object returned by `Connector.open()`. (See Chapter 7 for more information on service registration.)

Here are some examples of valid server connection strings and their meaning:

"btspp://localhost:102030405060708090A1B1C1D1E100;name=Print_Server;master=false" establishes a server connection with the UUID 0x102030405060708090A1B1C1D1E100 in the service record. The connection string also specifies that the ServiceName attribute is "Print_Server" and that the server can be either the master or the slave of the connection.

"btspp://localhost:1231242432434AAAABB;authenticate=true;author ize=true;name=Echo" establishes a server connection with the 0x1231242432434AAAABB UUID in the service record and the Service- Name attribute set to "Echo." All communication to the server must be authenticated and authorized.

"btspp://localhost:AB93248543812312312311ADEFE;encrypt=true; authorize=true;master=true" creates a server connection object with a service record that has the UUID 0xAB9324854381231231231- ADEFE in its service record. The server connection must be the master of the link. As far as security is concerned, the link must be authenticated, encrypted, and authorized. (Authentication is implied by setting the encrypt or authorize parameters to `true`.)

After `Connector.open()` returns a `StreamConnectionNotifier` object, we are ready to attempt to establish a connection. The `accept-AndOpen()` method should be called after `Connector.open()`. This method blocks until a client connects to the server. The `acceptAndOpen()` method returns a `StreamConnection` object. With the `Stream-Connection` object, the application can read and write to the client application.

To show how to create a simple RFCOMM application, we will develop an echo application. The `EchoServer` MIDlet accepts connec- tions from the `EchoClient` MIDlet, described later in this chapter. The `EchoServer` then reads messages sent from the client and sends the same message in reply. The `BluetoothMIDlet` class from earlier is reused. The thread started by the `BluetoothMIDlet` accepts connections from cli- ents. The `run()` method of this thread creates a `Form` and sets it to the current display. An "Exit" `Command` is added to the `Form` to destroy the MIDlet. Recall that the `BluetoothMIDlet` processes all `Command` events by destroying the MIDlet, which is exactly what we need it to do here.

```
package com.jabwt.book;
import java.lang.*
import java.io.*;
import javax.microedition.lcdui.*
import javax.microedition.io.*
import javax.bluetooth.*
public class EchoServer extends BluetoothMIDlet {
```

```
/**
 * Accepts connections from RFCOMM clients and
 * echoes back what is received from the client.
 * This method also displays the messages from a
 * client on a Form. It also displays on the Form the
 * connection string to use to connect to this service.
 */
public void run() {
  // Create the output Form and set it to be the
  // current Displayable
  Form msgForm = new Form("Echo Server");
  msgForm.addCommand(new Command("Exit",
    Command.EXIT, 1));
  msgForm.setCommandListener(this);
  Display.getDisplay(this).setCurrent(msgForm);
  }
}
```

Next, a `StreamConnectionNotifier` object must be created to accept connections from the client. After the notifier object is created, the `displayConnectionString()` method is called. This method determines the connection string that a client should use to connect to this server. This connection string is appended to the `Form`. The connection string is used by the client to eliminate the need to do device and service discovery. The changes needed to append the connection string to the `Form` are shown below. Throughout the book a gray box is used to identify additions or changes to code shown previously. Some of the code shown previously will be repeated to provide context, but this repeated code will appear outside the gray box.

```
public class EchoServer extends BluetoothMIDlet {
  ...

  /**
   * Adds the connection string to use to connect to
   * this service to the screen.
```

```
 *
 * @param f the Form to add the connection string to
 * @param notifier the notifier object to retrieve
 * the connection string from
 */
private void displayConnectionString(Form f,
  StreamConnectionNotifier notifier) {

  try {
    // Retrieve the connection string to use to
    // connect to this server
    LocalDevice device = LocalDevice.getLocalDevice();
    ServiceRecord record = device.getRecord(notifier);

    String connString = record.getConnectionURL(
      ServiceRecord.NOAUTHENTICATE_NOENCRYPT, false);

    int index = connString.indexOf(';');
    connString = connString.substring(0, index);

    // Display the connection string on the Form
    f.append(new StringItem("Connection String:",
      connString));
  } catch (BluetoothStateException e) {
    f.append("BluetoothStateException: " +
      e.getMessage());
  }
}
```

```
...
  public void run() {
    // Create the output Form and set it to be the
    // current Displayable
    Form msgForm = new Form("Echo Server");
    msgForm.addCommand(new Command("Exit", Command.EXIT, 1));
    msgForm.setCommandListener(this);
    Display.getDisplay(this).setCurrent(msgForm);
```

```
        try {
          // Create the notifier object
          StreamConnectionNotifier notifier =
            (StreamConnectionNotifier)
            Connector.open("btspp://localhost:"+
            "123456789ABCDE;name=Echo Server");

          //Display the connection string on the Form
          displayConnectionString(msgForm, notifier);
        } catch (IOException e) {
          msgForm.append("IOException: "+ e.getMessage());
        }
```

```
    }
  }
```

The final part of the `EchoServer` MIDlet is the most important. After the connection string is displayed on the `Form`, the `run()` method enters a forever loop that accepts connections from a client via a call to `acceptAndOpen()`. The input and output streams are opened once the connection has been established. The `run()` method then reads data from the `InputStream`. After the data is read, the `run()` method appends the data to the `Form` and sends the data in reply. The `run()` method continues to read data until the client closes the input stream.

```
public class EchoServer extends BluetoothMIDlet {
...

  public void run() {
    // Create the output Form and set it to be the
    // current Displayable
    Form msgForm = new Form("Echo Server");
    msgForm.addCommand(new Command("Exit",
      Command.EXIT, 1));
    msgForm.setCommandListener(this);
    Display.getDisplay(this).setCurrent(msgForm);
    try {
      // Create the notifier object
```

```
StreamConnectionNotifier notifier =
  (StreamConnectionNotifier)
  Connector.open(
  "btspp://localhost:123456789ABCDE;"
  + "name=Echo Server");

// Display the connection string on the Form
displayConnectionString(msgForm, notifier);

// Continue accepting connections until the MIDlet
// is destroyed
for (;;) {
  StreamConnection conn = notifier.acceptAndOpen();
  OutputStream output = conn.openOutputStream();
  InputStream input = conn.openInputStream();

  // Continued reading the input stream until the
  // stream is closed. Display the data on the
  // screen and write it to the output stream.
  byte [ ] data = new byte[10]
  int length =0;
  while ((length = input.read(data)) != -1) {
    msgForm.append(new String(data, 0, length));
    output.write(data, 0, length);
    output.flush();
  }
  // Close the streams and the connection
  output.close();
  input.close();
  conn.close();
}

} catch (IOException e) {
  msgForm.append("IOException: " + e.getMessage());
}
  }
}
```

4.3.2   Establishing a Client Connection

To establish a client connection, the `btspp` {scheme} is used. The {target} starts with two slashes followed by the Bluetooth address of the device to connect to and the server channel identifier of the service to connect to. The client connection string takes `master`, `authenticate`, and `encrypt` as {params}. When this connection string is passed to `Connector.open()`, the JABWT implementation attempts to establish a connection to the desired service. If the connection is established, `Connector.open()` returns a `StreamConnection` object, which allows the application to read and write to the server. Unlike the server's connection, the client's connection to the server has been established once `Connector.open()` returns.

What is the server channel identifier, and how does a service get one? The server channel identifier is similar to a TCP/IP port number. It uniquely identifies a service on a device. The server channel identifier is a number between 0 and 31. The server channel identifier is assigned by the JABWT implementation for a service. The server channel identifier is set in the service record's ProtocolDescriptorList attribute. This allows the `ServiceRecord's` `getConnectionURL()` method to generate the connection string to use to connect to the service. Because a device is not aware of the devices and services in an area, it is expected that most JABWT applications will use the `getConnectionURL()` method.

Now for some examples of client connection strings:

"btspp://008003DD8901:1;authenticate=true" creates an RFCOMM connection to the device with a Bluetooth address of 008003DD8901. It connects to the service identified by the server channel identifier 1. The connection string also causes the remote device to be authenticated.

"btspp://008012973FAE:5;master=true;encrypt=true" establishes an RFCOMM connection to the Bluetooth device with the address of 008012973FAE. The connection string connects to server channel 5. The connection string requires the local device to be the master of the connection and the link to be authenticated and encrypted.

After the connection is established and a `StreamConnection` object is obtained from `Connector.open()`, the input and output streams should be used to send and receive data. The streams are available via the `openInputStream()`, `openDataInputStream()`, `openOutput-Stream()`, and `openDataOutputStream()` methods. To end the connection, the `close()` method must be called on the `StreamConnection` object and any open input or output streams.

The `EchoClient` MIDlet shows how to establish an RFCOMM connection to a server and how to communicate with the server. The `EchoClient` allows a user to send messages to the `EchoServer` MIDlet, which echoes back what is sent. The `EchoClient` then reads the reply and appends the reply to a `Form` so that the user can see what was sent.

To eliminate the need to perform device and service discovery, the Bluetooth address and server channel is retrieved from the user via a `Form`. The user enters the Bluetooth address and server channel from the connection string displayed on the `EchoServer` screen when the `EchoServer` starts. Figure 4.8 shows the `EchoServer` and how it

Figure 4.8   `EchoServer` MIDlet when it starts (emulation only).

displays the connection string to use to connect to this server. In this example, the Bluetooth address to connect to is 002233445566, and the server channel is 1. When the EchoClient starts, a Form is displayed that asks the user to enter the Bluetooth address and server channel for the echo server. After the user enters the information for the server, the user can select the "Connect" Command.

```java
package com.jabwt.book;

import java.lang.*;
import java.io.*
import javax.bluetooth.*
import javax.microedition.io.*
import javax.microedition.lcdui.*;
import javax.microedition.midlet.*;
public class EchoClient extends BluetoothMIDlet {
  /**
   * The Form that interacts with the user. Used to
   * retrieve the connection information and the
   * text to send.
   */
  private Form connForm;

  /**
   * The Command used to Connect to the server.
   */
  private Command connectCommand;

  /**
   * Called when the MIDlet is made active. This
   * method displays a Form that retrieves the
   * Bluetooth address and the channel ID of the
   * Echo Server.
   */
  public void startApp() throws
    MIDletStateChangeException {
```

```
    // Create the Form. Add the Connect and Exit
    // Commands to the Form.
    connForm = new Form("Echo Client");
    connectCommand = new Command("Connect",
      Command.OK, 1);
    connForm.addCommand(connectCommand);
    connForm.addCommand(new Command("Exit",
      Command.EXIT, 1));
    connForm.setCommandListener(this);

    // Add the TextFields to retrieve the
    // Bluetooth address and channel
    // ID of the Echo Server
    TextField address = new TextField("Address",
      null, 12, TextField.ANY);
    connForm.append(address);

    TextField channel = new TextField("Channel",
      null, 2, TextField.NUMERIC);
    connForm.append(channel);

    Display.getDisplay(this).setCurrent(connForm);
  }
}
```

Now that the Bluetooth address and server channel have been retrieved, a connection must be made to the EchoServer. To make a connection, a new thread is created and started if the "Connect" Command is selected. This requires the EchoClient to implement the Runnable interface and define a run() method. The run() method creates the connection string and then attempts to establish a connection. The run() method also removes the two TextFields that retrieved the Bluetooth address and server channel ID. If a connection can be established, the "Connect" Command is replaced with the "Send" Command, and a TextField is added to the Form to request a message to send.

```
public class EchoClient extends BluetoothMIDlet {

    // The InputStream to receive data from the server.
    private InputStream input;

    // The OutputStream to send data to the server.
    private OutputStream output;

    // The connection to the server
    private StreamConnection conn;
```

...

```
    public void commandAction(Command c, Displayable d) {
        if (c.getCommandType() == Command.OK) {
            // The Connect Command was selected so start a
            // thread to establish the connection to the
            // server
            new Thread(this).start();
        } else {
            notifyDestroyed();
        }
    }
    /**
     * Create the connection string from the information
     * entered by the user.
     * @return the connection string
     */
    private String getConnectionString() {
        // Retrieve the TextFields from the Form
        TextField address = (TextField)connForm.get(0);
        TextField channel = (TextField)connForm.get(1);

        // Create the connection string
        StringBuffer temp = new StringBuffer("btspp://");
        temp.append(address.getString());
        temp.append(":");
        temp.append(channel.getString());
```

```
      // Remove the TextFields from the Form
      connForm.delete(0);
      connForm.delete(0);

      return temp.toString();
    }
    /**
     * Establishes a connection to the server.
     *
     * @param connString the connection string to connect
     * to the server
     * @return true if the connection was established;
     * false if the connection failed
     */
    private boolean connectToServer(String connString) {
      try {

        // Establish a connection to the server
        conn = (StreamConnection)
          Connector.open(connString);

        // Retrieve the input and output streams to
        //communicate with
        input = conn.openInputStream();
        output = conn.openOutputStream();

        return true;
      } catch (IOException e) {
        connForm.append("Connect failed (IOException: ");
        connForm.append(e.getMessage());
        connForm.append(")\n");
        return false;
      }
    }
    /**
     * Retrieves the Bluetooth address and channel ID
     * from the Form. This method then establishes a
     * connection to the server.
     */
```

```
public void run() {
  String connString = getConnectionString();
  connForm.append("Connecting to Server\n");
  if (connectToServer(connString)) {
    connForm.append("Done");

    // Remove the Connect Command and add the Send
    // Command to this Form
    connForm.removeCommand(connectCommand);
    Command sendCommand = new Command("Send",
      Command.SCREEN, 1);
    connForm.addCommand(sendCommand);

    // Add a TextField to the Form to retrieve the
    // text to send to the server from the user
    connForm.append(new TextField("Text to send",
      null, 20, TextField.ANY));
  }
}
```

}

Most of the previous code handles user interaction. The only code that uses JABWT is the `connectToServer()` method. The `connectTo-Server()` method establishes the connection and retrieves the input and output streams. The `getConnectionString()` method makes the `connectToServer()` method work because it specifies the `btspp` connection scheme, which specifies that the SPP and RFCOMM should be used to connect to the server.

The next step is to add code that sends a message to the server and reads the reply. To minimize the amount of work done within the MIDP `CommandListener` event handler, all of the communication with the server is done in a separate thread. To perform the processing in a separate thread, a new class must be created that implements the `Runnable` interface. The `Message` class does this. The `Message` class takes in its constructor the message, the input stream, and the output stream. When it starts, the thread of the `Message` class writes the message to the `OutputStream`. It then reads the reply from the server and displays it on the `Form` the user is currently viewing.

```
public class EchoClient extends BluetoothMIDlet {
    ...

    /**
     * Sends a message and reads the echo in reply.
     * Displays the reply on the screen and adds the
     * TextField to the end of the Form.
     */
    class Message implements Runnable {
        // The message to send to the server.
        private String theMessage;

        // The InputStream to read the reply from.
        private InputStream input;

        // The OutputStream to send the message to.
        private OutputStream output;

        /**
         * Creates a new Message to send to the server.
         *
         * @param msg the message to send
         * @param in the InputStream to read the reply from
         * @param out the stream to write the message to
         */
        public Message(String msg, InputStream in,
            OutputStream out) {
            theMessage = msg;
            input = in;
            output = out;
        }
```

```
/**
 * Sends the message to the server and reads the echo
 * in reply. The method adds the echo to the Form and
 * then adds a new TextField to the end of the Form.
 */
public void run() {
  try {
    // Send the message to the server.
    byte[] data = theMessage.getBytes();
    output.write(data);
    output.flush();

    // Read the reply and keep it in a StringBuffer
    // until the full reply is received.
    int fullLength = data.length;
    int length = input.read(data);
    fullLength -= length;
    StringBuffer buf = new StringBuffer(new
      String(data, 0, length));

    while (fullLength > 0) {
      length = input.read(data);
      fullLength -= length;
      buf = buf.append(new String(data, 0, length));
    }

    // Display the reply on the Form and remove the
    // final new line sent from the server
    connForm.append("\n");
    String displayString = buf.toString();
    displayString = displayString.substring(0,
      displayString.length() - 1);
    connForm.append(displayString);
  } catch (IOException e) {
    connForm.append("\nFailed to send message: " +
      e.getMessage());
  }
```

```
        connForm.append(new TextField("Text to send",
            null, 20, TextField.ANY));
    }
}
```

```
}
```

The final step is to use the `Message` class within the `EchoClient` MIDlet. This requires modifying the `commandAction()` method. The `if` statement is changed to a `switch` statement to determine whether the "Send," "Exit," or "Connect" `Command` was selected. If the "Send" `Command` was selected, the `commandAction()` method determines whether the last element in the `Form` is a `TextField`. This check is done to prevent two messages from being sent at the same time. The `TextField` is not the last `Item` if a message is currently being sent. If no message is being sent, then the `TextField` is the last `Item`. After this check is made, the `command-Action()` method creates a new `Message` object and starts the `Message` object in a thread. This `thread` sends the message and receives the reply.

```
public class EchoClient extends BluetoothMIDlet {
...
    public void commandAction(Command c, Displayable d) {
```

```
        switch (c.getCommandType()) {
        case Command.OK:
```

```
            // The Connect Command was selected so start a
            // thread to establish the connection to the server
            new Thread(this).start();
```

```
            break;
        case Command.SCREEN:
            // The Send Command was selected so send the
            // message to the server
```

```
          int index = connForm.size() - 1;

          // If the last Item is a TextField, then no message
          // is currently being sent so send a Message.
          Item item = connForm.get(index);
          if (item instanceof TextField) {
            TextField field = (TextField)item;
            connForm.delete(index);

            // Start a thread to send the message to the
            // server and process the reply
            new Thread(new Message(field.getString() +
              "\n", input, output)).start();
          }
          break;
        case Command.EXIT:

          // The Exit Command was selected so destroy the
          // MIDlet
          try {
            input.close();
            output.close();
            conn.close();
          } catch (Exception e) {
          }

          notifyDestroyed();

          break;
        }

      } ...
    }
```

This completes the echo client/server application. Figure 4.9 shows the `EchoClient` and `EchoServer` running. Now the `EchoClient` is able to send messages to the server while the `EchoServer` is able to

(A)   (B)

Figure 4.9 `EchoClient` (A) and `EchoServer` (B) communicating over RFCOMM (emulation only).

echo back any message sent from the client. Most of the code for both applications is not specific to JABWT but is MIDP code that provides for interaction between the application and the user. This is likely the case for most applications that use JABWT.

4.4 Summary

RFCOMM will likely be the most used Bluetooth protocol within JABWT because RFCOMM provides serial two-way communication and reuses familiar APIs from Java ME. The SPP is the Bluetooth profile realization of RFCOMM. Many Bluetooth profiles are built on the SPP to take advantage of existing serial port applications and protocols developed for wired communication.

Important concepts introduced in this chapter are links and connections. Two Bluetooth devices may have only a single Bluetooth link between them, but this link allows multiple Bluetooth connections. Although links are device by device, connections are at the JABWT application layer. In addition to the terms *link* and *connection,* the concepts of master and slave devices are introduced. The master device drives the frequency-hopping sequence used to communicate between two devices. Different Bluetooth profiles require one device to be the master and another device the slave. Being the master allows a device to accept and establish connections to other devices. By establishing these additional connections, the master device is able to set up a piconet.

The basic concepts of Bluetooth security are covered. Bluetooth provides four types of security on a link basis. Pairing is the initial process of identifying two devices to each other by exchanging a PIN outside of Bluetooth communication. Pairing sets up a shared secret between the devices so that pairing does not need to be completed every time. After pairing is completed, authentication can occur. Authentication is the process of verifying the identity of another device. After authentication has occurred, encryption and/or authorization can occur. Encryption is the process of encoding and decoding a message so that an eavesdropper cannot listen in on the conversation. Finally, authorization is the process of determining whether another device has permission to use a specific service.

Because RFCOMM provides reliable two-way communication, the `StreamConnection` and `StreamConnectionNotifier` interfaces from the GCF are reused. All RFCOMM connections start with a call to `Connector.open()` with a valid RFCOMM connection string. The connection string can include parameters for master/slave and Bluetooth security. If a client connection string is used in `Connector.open()`, a `StreamConnection` object is returned when the connection is established to the server. An RFCOMM server is created by calling

`Connector.open()` with a server connection string, and a `StreamConnectionNotifier` object is returned. With the `Stream-ConnectionNotifier` object, the server can accept connections from RFCOMM clients by calling `acceptAndOpen()`. After the connection has been established, input and output streams can be retrieved to read and write data.

5 OBEX

This chapter covers the following topics:

- What is OBEX?
- When should OBEX be used?
- How does the OBEX API fit into the JABWT specification?
- How does OBEX work?
- Establishing an OBEX connection
- Setting and retrieving OBEX headers
- Initiating and responding to OBEX requests
- Using OBEX authentication

5.1 Overview

The IrOBEX (Infrared Object Exchange protocol) [33] is defined by IrDA as an alternative to the HyperText Transport Protocol (HTTP) for embedded devices. IrOBEX targets memory-constrained embedded devices, which have slower processing speeds. Whereas HTTP makes a single request and a single reply, IrOBEX allows devices to break up requests and replies into smaller chunks. By breaking up the requests into smaller chunks of data, IrOBEX allows the data to be processed as it is received and allows a request or reply to be aborted.

IrOBEX, like HTTP, is transport neutral. In other words, IrOBEX works over almost any other transport layer protocol. Whereas the initial implementations of IrOBEX used infrared as the transport, there are presently implementations of IrOBEX that are running over TCP, serial, and RFCOMM connections. Because IrOBEX may run over different

transports and can break up requests and replies, IrOBEX may be opti-
mized to a specific transport protocol. What does this mean? Every
IrOBEX packet is segmented to fit within each transport layer packet.
This allows for efficient use of bandwidth.

IrOBEX has become even more popular since the Bluetooth SIG
licensed the protocol from IrDA. When the protocol is used with Bluetooth wireless technology, the *Ir* is dropped, and the protocol is referred
to as *OBEX*. (From this point forward, *OBEX* and *IrOBEX* are used inter-
changeably.) The Bluetooth SIG defined OBEX as one of the protocols in
the Bluetooth protocol stack. OBEX sits on the RFCOMM protocol. The
Bluetooth SIG went a step farther. The SIG realized that OBEX is an
excellent building block protocol from which to create Bluetooth profiles. To facilitate building new profiles, the Bluetooth SIG defined the
GOEP [11] to be the profile that defines how OBEX works within the
Bluetooth environment.

The OBEX API defined in JABWT is an optional API. This means
that the OBEX API may be implemented within a device that supports
the Bluetooth APIs, but just because a device supports the Bluetooth APIs
does not imply that it supports the OBEX APIs. Currently, most mobile
devices that support the Bluetooth APIs do not support the OBEX APIs.
In theory, the reverse is also possible—there could be support for the
OBEX API in devices that do not support the Bluetooth APIs. The reason
for this is that the OBEX API is independent of the Bluetooth APIs.

So why would a developer use OBEX on a device that has RFCOMM,
L2CAP, or TCP/IP? OBEX is a structured protocol that allows separation
of data and the attributes of data. Using OBEX allows clear definition of
one request from another. Using protocols such as RFCOMM or TCP/IP
requires the applications to know how data is sent and when to send the
reply. OBEX hides this within the protocol. OBEX is like the Extensible
Markup Language (XML). It provides structure to the data sent whereas
RFCOMM and TCP/IP simply send bytes.

5.1.1   Use Cases

OBEX can be used for a variety of purposes. The protocol is being used
in PDAs as a way to exchange electronic business cards. OBEX also has
been used to synchronize embedded devices with desktop computers.

The OBEX API defined for the Java programming language is intended to allow OBEX to be used for an even wider range of applications.

Synchronization

A common problem for MIDP devices, such as cell phones, is how to keep the information on the device synchronized with a desktop computer. With Bluetooth wireless technology, the device doesn't need to be connected to the PC and manually synchronized. Instead, the handset automatically synchronizes with the desktop PC when the device gets close to the PC. The Bluetooth SIG defined the Synchronization Profile to support such a use case. The Synchronization Profile utilizes OBEX to exchange data between the device and the PC.

Printing

Java ME has begun to be used by businesses as a way to keep in touch with employees. Being able to send and retrieve e-mail is now possible. Being able to update and check an employee's calendar and "to do" list also is possible. There is one drawback to using a Java ME device for these tasks. Most devices have a very limited screen size; therefore, users find it quite helpful for those devices to send e-mail or the calendar to a printer. Up to this point, the Java ME space contained devices that could talk back only with a central server. With the introduction of JABWT to Java ME, any two devices can talk. Sending documents to print is a natural use of OBEX. The Bluetooth SIG has released the Basic Printing Profile, which uses OBEX [31].

5.1.2 Protocol Description

OBEX is built on six basic operations: CONNECT, SETPATH, GET, PUT, ABORT, and DISCONNECT. The client initiates every operation with a request and waits for the server to send its response. Every OBEX session begins with a CONNECT request from the client to the server. (Although the IrOBEX specification defined a connectionless OBEX, it is not described here. The OBEX API defined by JABWT does not address this type of OBEX.) Every session ends with a DISCONNECT request. Between the CONNECT and DISCONNECT requests, the client may send any number of SETPATH, GET, ABORT, or PUT requests. The ABORT request is a special type of request. It ends a PUT or GET operation before the operation ends. (A PUT/GET operation is made of multiple PUT or GET requests and replies.)

Within each request and reply, OBEX headers may be sent. The OBEX specification defines a list of common headers. The common headers include but are not limited to

- NAME, which specifies the name of the object
- LENGTH, which specifies the length of the object
- TYPE, which specifies the Multipurpose Internet Mail Extensions (MIME) type of the object
- COUNT, which is used by a CONNECT request to specify the number of objects to send or receive
- DESCRIPTION, a short description of the object
- HTTP, which specifies an HTTP header
- BODY, which specifies part of the object
- END OF BODY, which specifies the last part of the object

The OBEX specification defines how these common headers are encoded. For example, the NAME header must be a Unicode string. The BODY and END OF BODY headers are used to send or retrieve objects from a server via PUT or GET requests. The END OF BODY signals to the receiver that this is the last chunk of the object. In addition to the common headers, the specification also allows 64 user-defined headers. The specification breaks these headers into four groups of 16 headers. Each group represents a different type of data. There are groups for Unicode strings, 4-byte unsigned integers, single bytes, and byte sequences.

The OBEX specification defines two additional special operations: the PUT-DELETE and CREATE-EMPTY operations. The PUT-DELETE operation is a PUT operation with a NAME header and no BODY header. This operation is used to tell the server to delete the object with the specified name. The CREATE-EMPTY operation also is a PUT operation, but the CREATE-EMPTY operation contains a NAME and an END OF BODY header with no data. The CREATE-EMPTY operation signals to the server to create the object with the specified name with nothing in the object.

5.1.3 Example Session

Every OBEX session begins with the client issuing a CONNECT request. If the client wants, the client can include additional headers to send to the server. When the server receives the request, the server processes the headers and decides whether it will accept the connection request. If the server accepts the request, the server responds with an OK, SUCCESS response code. If the server rejects the request, the server responds with one of the HTTP response codes that specify why the request was not accepted. In the example in Figure 5.1, the client issues the CONNECT request and sends the COUNT header to the server to specify the number of objects to be transferred. The server processes the request and replies with the OK, SUCCESS response code.

After the connection is established, the client may want to change to a different location on the server. The client is able to change folders by using the SETPATH operation. In Figure 5.2, the client sends the SETPATH operation and specifies the name of the directory to change to by using the NAME header. When the server receives the request, it may decide to allow or not allow the change. The server can deny the request for a variety of reasons, including using the NOT FOUND response if the folder does not exist on the server.

Even though the server was not able to fulfill the SETPATH operation, the session is still active and the client may continue to make requests to the server. For example, the client may want to send a file

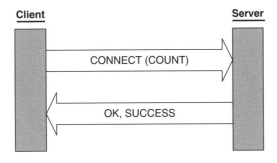

Figure 5.1　OBEX CONNECT operation.

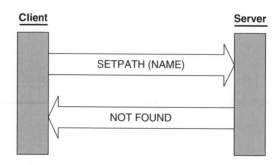

Figure 5.2   OBEX SETPATH operation.

to the server. To do this, the client issues a PUT request. If the file is large, the client may need to break the file up into smaller chunks to send to the server. If this is the case, the client sends the initial PUT request with a NAME header, to specify the name of the file, and the BODY header containing the first chunk of the file. When the server receives this request, the server stores the first chunk of the file and replies with a CONTINUE response. When the client receives the CONTINUE response, the client sends the next chunk of the file via another PUT request with the BODY header. After storing this part of the file, the server sends another CONTINUE response. This back and forth continues until the last chunk of data is sent to the server (see Figure 5.3). For the last chunk of the file, the client again sends a PUT request, but the client includes the data in an END OF BODY header rather than a BODY header. This header signals to the server that this is the last piece of the file. After the server receives notice from the client that no more data will be sent, the server responds with an OK, SUCCESS response code. When the client receives this response code, the client knows that the object was successfully sent to the server.

To end the OBEX session, the client must issue a DISCONNECT request. Usually, a DISCONNECT request does not contain any additional headers, but OBEX does not restrict headers from being included in the DISCONNECT request. When it receives a DISCONNECT request, the server frees any resources that it may have allocated and sends an OK, SUCCESS response to the client (see Figure 5.4). When the client receives this response, the OBEX session has ended.

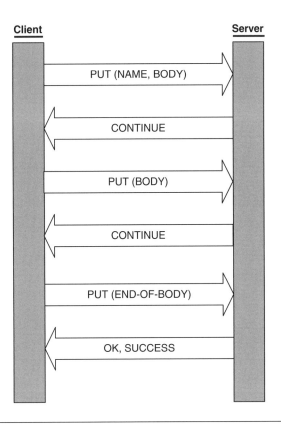

Figure 5.3   OBEX PUT operation.

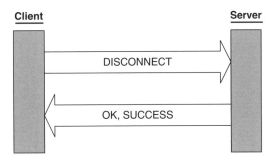

Figure 5.4   OBEX DISCONNECT operation.

It is important to remember that when the OBEX session has ended, the physical connection between the two devices may not have been closed. The transport protocol also must be disconnected. How or when this is done is not specified by the OBEX specification.

5.2 API Capabilities

The OBEX API is quite different from most existing OBEX implementations, which provide only a high-level interface to the protocol. For example, Palm developers can use an API for OBEX that allows a user to send a business card or receive an appointment but not to control how the data was sent. Control of the specifics of the transaction is not available. The Java OBEX API provides a low-level interface. The low-level interface gives developers more control over each request and reply, adding a new layer of complexity.

Although the OBEX API provides greater access to the protocol, the OBEX API hides some of the details of the protocol from developers. The OBEX API handles all the translation of OBEX headers to their corresponding byte representation. The API also hides some of the details of the CONNECT request. For example, the OBEX API implementation handles the negotiation of OBEX packet sizes. Because the packet size is not available to an application developer, the OBEX API implementation handles converting requests into multiple packets for PUT and GET requests. This allows an application to simply send the BODY data while relying on implementation of the API to convert the BODY data into different packets.

To make it easier to learn, the OBEX API was based on other Java APIs with which many developers are familiar. The client API is designed from the combination of the `javax.microedition.io.Content-Connection` interface and the `javax.microedition.io.Datagram-Connection` interface from the GCF (see Figure 5.5). GET, PUT, and CREATE-EMPTY operations use the `javax.obex.Operation` interface, which extends the `ContentConnection` interface. The CONNECT, SETPATH, PUT-DELETE, and DISCONNECT operations work as the `DatagramConnection` interface does. For sending a message with the `DatagramConnection`, a `javax.microedition.io.Datagram` object must be created and used as the argument to the `send()` method of the

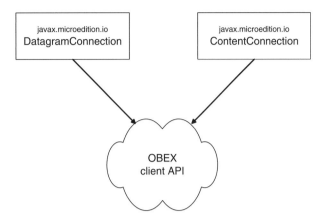

Figure 5.5   OBEX client API resulted from the combination of two connection types.

`DatagramConnection` interface. Similarly, for sending OBEX headers, a `javax.obex.HeaderSet` object must be created and passed to the `connect()`, `setPath()`, `delete()`, and `disconnect()` methods of the `javax.obex.ClientSession` interface.

For an OBEX server, the OBEX API combines concepts from the `javax.microedition.io.StreamConnectionNotifier` interface and the Java servlet API (see Figure 5.6). The server API, like the client API,

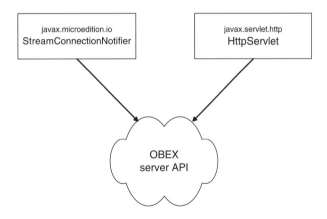

Figure 5.6   OBEX server API was created by combining two well known APIs.

is based on the GCF. After a `SessionNotifier` object is created by calling `Connector.open()`, `acceptAndOpen()` is called with a `javax.obex.ServerRequestHandler` object. The `ServerRequestHandler` class is similar to the `java.servlet.http.HttpServlet` class. The `Server-RequestHandler` class defines methods for each type of OBEX request that a server may receive, such as `onConnect()`, `onDisconnect()`, `onPut()`, `onGet()`, `onDelete()`, and `onSetPath()`. Only requests to which a server wants to respond must be implemented.

The OBEX API also provides a mechanism for OBEX authentication. OBEX authentication works via a challenge and response scheme using two OBEX headers. The AUTHENTICATION_CHALLENGE header is sent when an application on one device wants to authenticate an application on another device. When a device receives an AUTHENTICATION_CHALLENGE header, it combines the shared secret or password with the 16-byte challenge received in the AUTHENTICATION_CHALLENGE header. The Message Digest 5 (MD5) hash algorithm is applied to the combined password and challenge. The resulting value is returned in an AUTHENTICATION_RESPONSE header. When the challenger receives the AUTHENTICATION_RESPONSE header, the challenger combines the 16-byte challenge sent in the original AUTHENTICATION_CHALLENGE header and the shared secret and applies the MD5 hash algorithm. The resulting value is compared with the value received in the AUTHENTICATION_RESPONSE header. If the two values are equal, the other device is authenticated.

OBEX authentication is different from Bluetooth authentication. Bluetooth authentication authenticates two Bluetooth devices to each other. OBEX authentication authenticates two users or applications to each other. Although Bluetooth authentication is handled at the Bluetooth stack and radio layer, OBEX authentication is handled at the application layer. OBEX authentication and Bluetooth authentication can be used at the same time.

The OBEX API uses an API similar to the Java SE authentication API for OBEX authentication. The OBEX API defines the `javax.obex.Authenticator` interface. When an AUTHENTICATION_CHALLENGE header is received, the `onAuthenticationChallenge()` method is called. This method returns a `javax.obex.PasswordAuthentication` object with the user name and password pair that will be used in creating the AUTHENTICATION_RESPONSE. When an AUTHENTICATION_RESPONSE header is received, the `onAuthenticationResponse( )` method is called. The shared secret or password is returned from the `onAuthenticationResponse()` method. The OBEX API implementation handles all the hashing of challenges/passwords and validation of the authentication request.

The OBEX API is based on the IrOBEX 1.2 version of the specification. There have been updates to the IrOBEX specification that are not backwards compatible. In particular, values of some headers have been changed. These changes have not been made to the OBEX API defined in JSR-82.

5.3 Programming with the API

The OBEX API is built on the GCF defined in CLDC. The OBEX API adds three new interfaces that extend the `javax.microedition.io.Connection` interface. The `javax.obex.ClientSession` interface is returned from `Connector.open()` when a client connection string is provided. The `javax.obex.SessionNotifier` interface is returned from the `Connector.open()` method for server connections. Finally, the `javax.obex.Operation` interface is used to process PUT and GET requests. The `javax.obex.Operation` interface hides the back and forth nature of the PUT and GET requests (Figure 5.7).

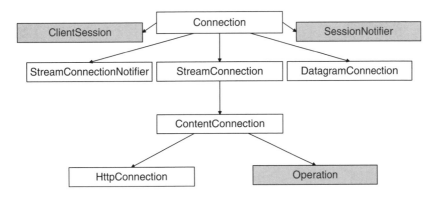

Figure 5.7   GCF with OBEX API.

In addition to these new interfaces, the OBEX API defines the `javax.obex.Authenticator` and `javax.obex.HeaderSet` interfaces. The `Authenticator` interface is implemented by applications that want to handle authentication challenges and responses (OBEX authentication is fully explained in Section 5.3.5). The `HeaderSet` interface encapsulates a set of OBEX headers. All OBEX headers except for the BODY, END-OF-BODY, CONNECTION-ID, AUTHENTICATION_CHALLENGE, and AUTHENTICATION_RESPONSE headers can be set in a `HeaderSet` object. OBEX headers not set within the `HeaderSet` interface can be set and retrieved by other methods.

The OBEX API introduces three new classes. The `javax.obex.PasswordAuthentication` class keeps user name and password pairs for OBEX authentication. The `javax.obex.ResponseCodes` class defines all the valid response codes that a server may send to a client. Finally, servers extend the `javax.obex.ServerRequestHandler` class. This class defines the methods called when the server receives different OBEX requests.

5.3.1   Establishing a Connection

For a client or a server to use the OBEX API, the client or server must first provide a connection string to `Connector.open()`. The OBEX API uses the same connection URL definition as the CLDC specification:

> {scheme}:[{target}][{params}]

The OBEX connection string is slightly different from the connection strings defined in MIDP and MIDP 2.0. Because OBEX can be used with a number of different transports, the connection string needs to specify the transport protocol in addition to specifying OBEX. The transport protocol is specified within the {scheme}. With the exception of OBEX over RFCOMM, the connection string defined by the OBEX API is

> {transport}obex://{target}{params}

If TCP/IP is the transport protocol used for an OBEX connection, the {scheme} is `tcpobex`. When opening a `tcpobex` client connection to a server, the {target} is the IP address and port number of the server. When opening a server connection, the {target} is just the port number of the server. In the TCP/IP case, there are no {params} defined for a client or server connection.

If RFCOMM is the transport protocol, the connection string does not follow this rule. This is because the GOEP is the realization of OBEX in the Bluetooth specification; therefore the {scheme} for OBEX over RFCOMM connections is `btgoep`. The {target} is the Bluetooth address and RFCOMM channel number to establish a client connection. For server connections, the {target} is the UUID of the service. All the valid {params} for RFCOMM are valid for OBEX over RFCOMM (see Table 4.1 for the list of valid {params} for RFCOMM).

Some example client connection strings are

1. btgoep://00802d5b12af:1;authenticate=yes
2. tcpobex://163.10.70.75:1505
3. irdaobex://discover;ias=MyAppOBEX,OBEX,OBEX:IrXfer;

See the JAWBT specification [22] for an explanation of the connection string for OBEX over IRDA.

Some server connection strings are

1. btgoep://localhost:1233212ADBAA9324BAFE23331231222C
2. tcpobex://:1801
3. irdaobex://localhost.0200

After `Connector.open()` is called with a client connection string, a `ClientSession` object is returned. A transport connection is established by a call to `Connector.open()`, but an OBEX layer connection has not yet been established. To establish an OBEX layer connection, `ClientSession.connect()` must be called. Before the transport layer connection is closed, `ClientSession.disconnect()` must be called to close the OBEX layer connection.

On the server side, the `SessionNotifier` object returned by `Connector.open()` is used to accept connections from clients by calling `acceptAndOpen()` on the `SessionNotifier` object. The `acceptAndOpen()` method takes a `ServerRequestHandler` argument and an optional `Authenticator` argument. A developer creates a new class that extends the `ServerRequestHandler` class and implements the methods for the type of requests the developer would like the server to handle.

For example, `onConnect()` should be implemented for CONNECT requests and `onGet()` for GET requests. The call to `acceptAndOpen()` does not return until a client connects. The `acceptAndOpen()` method returns a `Connection` object representing the transport layer connection to the client.

5.3.2 Manipulating OBEX Headers

OBEX communicates all of its information within headers. JABWT allows headers to be written and read via different methods based on the header. Of all the valid headers, the BODY, END-OF-BODY, AUTHENTICATION_ CHALLENGE, AUTHENTICATION_RESPONSE, and CONNECTION-ID headers have specific methods that allow developers to access them. All other headers can be accessed through the `HeaderSet` interface.

Developers are not allowed to define their own implementation for the `HeaderSet` interface. Instead, developers use implementations of the interface found within the API implementation. OBEX clients use the `createHeaderSet()` method defined in the `ClientSession` interface. On the other hand, OBEX servers are passed `HeaderSet` objects when they override an `onXXX()` method in the `ServerRequestHandler` class (see Section 5.3.4 for more information on how to implement an OBEX server).

Once a `HeaderSet` object is created or received, it is very easy to access different headers. Within the `HeaderSet` interface are constants defined for most of the headers in the OBEX specification. In addition to these constants are 64 user-defined headers that can be used. To set a header in the object, call the `setHeader()` method with the header identifier and the header's value. The header's value must be of the type specified in the OBEX API. Table 5.1 is the full list of headers that can be set with `setHeader()`, their meaning according to the OBEX specification, and the type of object to use. For example, the COUNT header must be set with a `java.lang.Long` object, and the NAME header must be set with a `java.lang.String`. If `setHeader()` is called with a different type, `IllegalArgumentException` is thrown. Likewise, to retrieve a header, use the `getHeader()` method with the header identifier. The `getHeader()` method also returns an object of the type specified in Table 5.1.

Table 5.1 OBEX Header Constants in the `HeaderSet` Interface, Their Meaning, and Their Type

Value	Meaning	Type
`COUNT`	Used by CONNECT to specify the number of objects to be communicated during the session	`java.lang.Long`
`NAME`	Name of the object	`java.lang.String`
`TYPE`	MIME type of the object	`java.lang.String`
`LENGTH`	Size of the object	`java.lang.Long`
`TIME_ISO_8601`	Time stamp of the object (recommended header to use to time stamp an object)	`java.util.Calendar`
`TIME_4_BYTE`	Time stamp of the object	`java.util.Calendar`
`DESCRIPTION`	Brief description of the object	`java.lang.String`
`TARGET`	Target OBEX service	`byte[]`
`HTTP`	Specifies an HTTP header	`byte[]`
`WHO`	OBEX service processing the request	`byte[]`
`OBJECT_CLASS`	OBEX object class of the object	`byte[]`
`APPLICATION_PARAMETER`	Application-specific parameter	`byte[]`
48 to 63 (0x30 to 0x3F)	User-defined headers to send a string	`java.lang.String`
112 to 127 (0x70 to 0x7F)	User-defined headers to send a byte array	`byte[]`
176 to 191 (0xB0 to 0xBF)	User-defined headers to send a byte	`java.lang.Byte`
240 to 255 (0xF0 to 0xFF)	User-defined headers to send an unsigned integer in the range of 0 to $2^{32} - 1$	`java.lang.Long`

Although some headers, such as NAME and COUNT, have a specific meaning, 64 headers are defined in OBEX that have no general meaning according to the OBEX specification. These are the user-defined headers. These headers should be used by applications to exchange data if the data does not fall into one of the defined OBEX headers.

The `HeaderSet` interface also provides a `getHeaderList()` method. This method returns an array of integers that represent the header identifiers set within the `HeaderSet` object. The `getHeaderList()` method never returns `null`. If no headers are available via the `getHeaders()` method, `getHeaderlist()` returns an empty array.

This method allows a developer to find all the headers included in a request or a reply without calling `getHeader()` on every header specified in the `HeaderSet` interface.

Five OBEX headers are handled differently. The BODY and END OF BODY headers are manipulated via input and output streams from an `Operation` object. The AUTHENTICATION_CHALLENGE and AUTHENTICATION_RESPONSE headers are accessed via the `Authenticator` interface. The CONNECTION-ID header can be retrieved and set through the `getConnectionID()` and `setConnectionID()` methods of `ClientSession` and `ServerRequestHandler`.

The CONNECTION-ID header is unique within OBEX. The CONNECTION-ID header is used to differentiate multiple services provided by a single OBEX notifier object. If the CONNECTION-ID header is set in the OBEX API, the header is included in every packet sent by the API implementation.

5.3.3 Sending a Request to the Server

After establishing a transport layer connection to a server through `Connector.open()`, the client must first issue a CONNECT request to the server to establish the OBEX layer connection. The client sends a CONNECT request by calling `connect()`. Within the CONNECT request, the client may include any headers by passing the headers to `connect()`. A `HeaderSet` object is returned from `connect()`. This `HeaderSet` object allows the client to get the headers received from the server and the response code. To access the response code sent by the server, the client calls the `getResponseCode()` method. The `getResponseCode()` method returns one of the response codes defined in the `ResponseCodes` class. If the server responds with `OBEX_HTTP_OK`, the OBEX layer connection has been established. The server can send headers in the response.

The following code establishes a transport layer connection to the server and then an OBEX connection. As part of the OBEX CONNECT request, the `COUNT` header and a user-defined header are sent. Next the `connectToServer()` method verifies that the connection has been accepted. If the server denies the connection, the `connectToServer()` method retrieves the `DESCRIPTION` header to find out the reason for the failure.

```
ClientSession connectToServer(String connString) throws
  IOException {
  // Establish the transport layer connection
  ClientSession conn =
    (ClientSession)Connector.open(connString);
  // Create the HeaderSet object to send to the server
  HeaderSet header = conn.createHeaderSet();
  // Set the headers to send to the server
  header.setHeader(HeaderSet.COUNT, new Long(3));
  header.setHeader(0x30, "New OBEX Connection");
  HeaderSet response = conn.connect(header);
  // Verify that the server accepted the connection
  if (response.getResponseCode() !=
    ResponseCodes.OBEX_HTTP_OK) {
    try {
      conn.close();
    } catch (Exception e) {
    }
    // The connection was rejected by the server so
    // throw an IOException.
    throw new IOException("Connection rejected (0x" +
      Integer.toHexString(response.getResponseCode()) +
      ": " + (String)response.getHeader(
      HeaderSet.DESCRIPTION) +
      ")");
  }
  return conn;
}
```

The `delete()` and `disconnect()` methods work in a similar way to `connect()`. The `setPath()` method works slightly differently. In a SETPATH request, the NAME header is used to specify to which directory to change to. In addition to passing in the NAME header in the `Header-Set` argument, any additional headers can be used. The `setPath()` method also takes two boolean arguments. The first argument is set to `true` if the server should move up one directory before moving to the directory specified by NAME. (This is similar to a cd.. in DOS.) The second argument, create, is set to true if the directory should be created if it does not exist. If the create argument is set to false, an error should occur if the client tries to move to a directory that does not exist. The following code moves to the directory specified by folderName.

```
void moveToDirectory(ClientSession conn,
  String folderName) throws IOException {
  // Specify the directory to move to
  HeaderSet header = conn.createHeaderSet();
  header.setHeader(HeaderSet.NAME, folderName);

  // Change to the directory specified. Do not backup
  // one directory (second argument) and do not create
  // it if it does not exist (third argument).
  HeaderSet reply = conn.setPath(header, false, false);

  // Validate that the server moved to the specified
  // directory
  switch (reply.getResponseCode()) {
  case ResponseCodes.OBEX_HTTP_OK:
    // The request succeeded so simply return from this
    // method
    return;
  case ResponseCodes.OBEX_HTTP_NOT_FOUND:
    // There was no directory with the name so throw an
    // IOException
    throw new IOException("Invalid directory");
  default:
    // The request failed for some other reason, so
    // throw a generic IOException
    throw new IOException(
      "Move to directory request failed");
  }
}
```

The PUT and GET operations work differently. Because PUT and GET requests pass body data between client and server, the put() and get() methods return an Operation object. To retrieve body data, open the

InputStream or DataInputStream by using the openInputStream() or openDataInputStream() methods, respectively. On the other hand, the OutputStream and DataOutputStream returned by openOutputStream() and openDataOutputStream() allow a client to send body data to the server. The OBEX implementation converts the BODY and non-BODY data headers to and from packets (see Figure 5.8).

Sending and retrieving data must follow a set of rules depending on the type of OBEX request. Even though multiple packets can be exchanged, PUT and GET operations are still broken into requests and responses. During the request portion of the PUT and GET operation, the operation may write to the OutputStream or DataOutputStream. During the PUT or GET response, BODY data may be read from the InputStream or DataInputStream.

For PUT requests, closing the OutputStream or DataOutput-Stream ends the request portion of the operation (see Figure 5.9). Calling getResponseCode() also causes the OutputStream to close and thus ends the PUT request and starts the response portion of the operation. It should be noted that calling read() on the InputStream before closing the OutputStream or before calling getResponseCode() causes the application to hang because the BODY data will not be available until the response portion of the operation. The response portion starts when the OutputStream is closed or getResponseCode() is called.

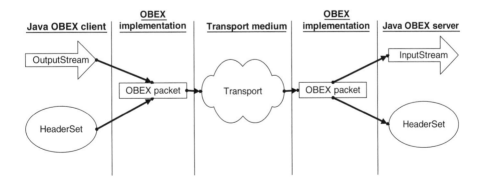

Figure 5.8 PUT request that combines OutputStream and HeaderSet into an OBEX packet.

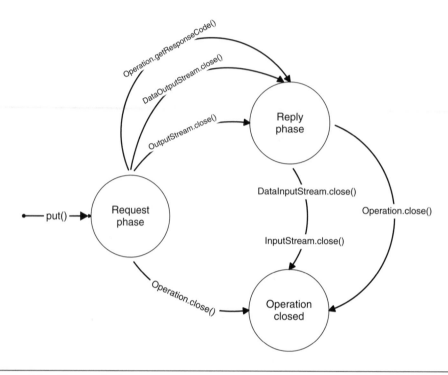

Figure 5.9   Phases of a PUT request.

The following code is an example of a PUT operation that sends an object to the server. The code also sends the TYPE and LENGTH headers in the PUT request.

```
void sendBytes(ClientSession conn, String type,
  byte[] data) throws IOException {
  // Set the headers in the HeaderSet to send to the server
  HeaderSet header = conn.createHeaderSet();
  header.setHeader(HeaderSet.TYPE, type);
  header.setHeader(HeaderSet.LENGTH, new
    Long(data.length));

  // Issue the PUT request to the server
  Operation op = conn.put(header);
```

```
  // Send the BODY data to the server
  OutputStream out = op.openOutputStream();
  out.write(data);
  out.close();

  // Verify that the server accepted the object
  if (op.getResponseCode() != ResponseCodes.OBEX_HTTP_OK) {
    op.close();
    throw new IOException("Request failed");
  }

  op.close();
}
```

GET operations work slightly differently from PUT operations. For GET operations, a call to openInputStream( ) or openDataInput-Stream( ) causes the request portion to end (see Figure 5.10). If get-ResponseCode( ) is called during a GET operation, the InputStream is closed, and no further BODY data can be read. Therefore do not call getResponseCode() until all the BODY data sent by the server is read.

The following method retrieves an object from the server using a GET operation.

```
byte[] getBytes(ClientSession conn, String name) throws
  IOException {
  // Create the request to send to the server
  HeaderSet header = conn.createHeaderSet();
  header.setHeader(HeaderSet.NAME, name);

  // Send the request to the server
  Operation op = conn.get(header);

  // Retrieve the bytes from the server
  InputStream input = op.openInputStream();

  // Read the data from the server until the end of
  //stream is reached
  ByteArrayOutputStream out = new ByteArrayOutputStream();
  int data = input.read();
  while (data != -1) {
    out.write(data);
    data = input.read();
```

```
    }
    input.close();

    // Verify that the whole object was received
    int responseCode = op.getResponseCode();
    op.close();

    switch(responseCode) {
    case ResponseCodes.OBEX_HTTP_OK:
      return out.toByteArray();
    case ResponseCodes.OBEX_HTTP_NOT_FOUND:
    case ResponseCodes.OBEX_HTTP_NO_CONTENT:
      // Since nothing was found, return null
      return null;
    default:
      throw new IOException("Request Failed");
    }
  }
```

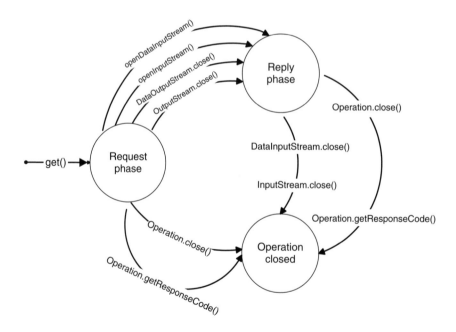

Figure 5.10   Phases of a GET operation.

A client can end a PUT and GET request during the operation by calling the `abort()` method. The `abort()` method sends an ABORT request to the server and signals to the server the request should not be processed. The `abort()` method closes the `InputStream`, `OutputStream`, and `Operation` object. If the operation has already ended, the `abort()` method throws an `IOException`.

5.3.4 Receiving a Request from a Client

OBEX servers are similar to Java servlets once the transport layer connection is established. HTTP servlets extend `HttpServlet`; OBEX servers extend the `ServerRequestHandler` class. Unlike `HttpServlet`, the `ServerRequestHandler` class does not require any methods to be overridden. OBEX servers need only to override the methods for the type of client requests the server would like to handle. For example, if the OBEX server wants only to process CONNECT, SETPATH, and PUT requests, the server needs only to override the `onConnect()`, `onSet-Path()`, and `onPut()` methods. If the client makes a DELETE or GET request, the implementation on the server side would automatically respond with the `OBEX_HTTP_NOT_IMPLEMENTED` response code. Table 5.2 shows how OBEX requests are received by an application and how the implementation will respond if the method is not overridden.

The `onConnect()`, `onDelete()`, and `onDisconnect()` methods allow servers to respond to OBEX CONNECT, DELETE, and DISCONNECT requests, respectively. All three methods have two arguments. The

Table 5.2 How OBEX Requests Are Translated to Methods by the JABWT Implementation

OBEX Request	`ServerRequestHandler` **Method**	Default Return Value
CONNECT	`onConnect()`	`OBEX_HTTP_OK`
SETPATH	`onSetPath()`	`OBEX_HTTP_NOT_IMPLEMENTED`
GET	`onGet()`	`OBEX_HTTP_NOT_IMPLEMENTED`
PUT	`onPut()`	`OBEX_HTTP_NOT_IMPLEMENTED`
DELETE	`onDelete()`	`OBEX_HTTP_NOT_IMPLEMENTED`
DISCONNECT	`onDisconnect()`	`OBEX_HTTP_OK`

first argument provides the headers sent from the client. The second argument provides the `HeaderSet` to set headers in the reply. The `onConnect()` and `onDelete()` methods both return the response code to send in the reply as specified in the `ResponseCodes` class. The `onDisconnect()` method returns nothing because a DISCONNECT request must send an `OBEX_HTTP_OK` response.

The following example code processes an OBEX CONNECT request. First, the COUNT and 0x30 user-defined headers are retrieved. If they exist in the request, the connection is accepted; otherwise the connection is rejected. If the connection is rejected, the description header specifies the cause of the rejection.

```
public int onConnect(HeaderSet request, HeaderSet reply) {
  try {
    // Retrieve the expected headers
    Long count = (Long)request.getHeader(HeaderSet.COUNT);
    String conName = (String)request.getHeader(0x30);
    if ((count == null) || (conName == null)) {
      reply.setHeader(HeaderSet.DESCRIPTION,
        "Required headers missing");
      return ResponseCodes.OBEX_HTTP_BAD_REQUEST;
    }

    return ResponseCodes.OBEX_HTTP_OK;
  } catch (IOException e) {
    reply.setHeader(HeaderSet.DESCRIPTION,
      "IOException: " + e.getMessage());
    return ResponseCodes.OBEX_HTTP_INTERNAL_ERROR;
  }
}
```

The `onSetPath()` method works similarly to `onConnect()`, `onDisconnect()`, and `onDelete()` methods with one exception. As part of a SETPATH request, the client may specify creation of the directory if it does not exist and backing up one directory before moving to the directory specified by the NAME header. The `onSetPath()` method has two boolean arguments to pass these values from the client's request to the server.

The onGet() and onPut() methods work differently. These methods have only a single argument, an Operation object. The Operation object provides access to the BODY header through the InputStream and OutputStream. Unlike Operation objects returned by the client's get() and put(), Operation objects received via the server's onGet() and onPut() methods do not have any special ordering rules. (For example, an Operation object received as an argument in onGet() may read from the InputStream first or it may write to the OutputStream first.) The Operation getResponseCode() and abort() methods throw an IOException if called on the server. If the server receives an ABORT from the client, the Operation is closed, and all the methods on the closed Operation throw an IOException.

In addition to BODY headers, GET and PUT operations can use additional headers. These headers can be retrieved via the getReceivedHeaders() method. This method returns a HeaderSet object containing all the latest headers received. Because GET and PUT operations may require multiple request and reply packets within a single operation, the same header may be sent multiple times. The getReceivedHeaders() method returns the HeaderSet object containing only the latest headers received. To send headers, a HeaderSet object must be created by calling the ServerRequestHandler.createHeaderSet() method. After the HeaderSet object is created, all headers that will be sent in the reply should be set. After the values are set, the headers are sent when the HeaderSet object is passed to the sendHeaders() method.

The following code reads the data sent to the server via a PUT request and stores it in a Record Management System (RMS) RecordStore.

```
public int onPut(Operation op) {
  int response = ResponseCodes.OBEX_HTTP_OK;

  try {
    // Retrieve the NAME header
    HeaderSet headers = op.getReceivedHeaders();
    String name =
      (String)headers.getHeader(HeaderSet.NAME);

    // Read the data from the input stream
    ByteArrayOutputStream out = new
      ByteArrayOutputStream();
```

```
    InputStream in = op.openInputStream();
    byte[] data = new byte[100];
    int length = in.read(data);
    while (length != -1) {
      out.write(data, 0, length);
      length = in.read(data);
    }
    in.close();
    // Open the RecordStore with the name from the NAME
    // header
    RecordStore store = RecordStore.openRecordStore(name,
      true);
    data = out.toByteArray();
    store.addRecord(data, 0, data.length);
    // Close the RecordStore
    store.closeRecordStore();
  } catch (Exception e) {
    HeaderSet header = createHeaderSet();
    header.setHeader(HeaderSet.DESCRIPTION,
      e.getMessage());
    try {
      op.sendHeaders(header);
    } catch (Exception ex) {
    }
    response = ResponseCodes.OBEX_HTTP_INTERNAL_ERROR;
  }
  // Close the Operation
  try {
    op.close();
  } catch (Exception e) {
  }
  return response;
}
```

The OBEX API implementation verifies that the response code is valid before it is sent to the client. The implementation changes the response code to OBEX_HTTP_INTERNAL_ERROR if the onXXX() method returns something other than a response code specified in the `Response-Codes` class. If an uncaught exception is received by the implementation, the implementation changes the response code to `OBEX_HTTP_INTERNAL_ERROR`.

5.3.5 Using OBEX Authentication

OBEX authentication works via a challenge and response mechanism. To authenticate the other end of an OBEX connection, an AUTHENTICATION_CHALLENGE header is sent with a challenge. To respond to an authentication request, an AUTHENTICATION_RESPONSE header is sent with a hash of the challenge and password and an optional user name. Even though authentication usually occurs during a CONNECT request, OBEX authentication can occur at any time during an OBEX session. Before OBEX authentication is used, an `Authenticator` must be created and set via a call to `Client-Session.setAuthenticator()` or `SessionNotifer.acceptAndOpen()`. If no `Authenticator` is specified for a client or server, any authentication requests or replies will fail.

To send an AUTHENTICATION_CHALLENGE header, simply call the `createAuthenticationChallenge()` method on a `HeaderSet` object that will be used in a request or a reply. The `createAuthenticationChallenge()` method allows a developer to specify which user name and password to include via a description parameter. The method also allows the developer to specify whether a user name is required (second argument) and whether full access will be granted if the authentication succeeds (third argument).

The following code is an example of a client sending an authentication challenge to the server. It sets the `Authenticator` and issues an authentication challenge within a CONNECT request. The following code must set the `Authenticator` to handle the authentication response header from the server. (The `ClientAuthenticator` class is defined later in this chapter.)

```
ClientSession connectToServer(String connString) throws
   IOException {
   // Create the transport layer connection to the server
   ClientSession conn = (ClientSession)
```

```
Connector.open(connString);
// Set the AUTHENTICATION_CHALLENGE header to send to
// the server. The second argument to
// createAuthenticationChallenge() specifies that a
// user name is required. The third argument specifies
// whether full access will be granted.
HeaderSet request = conn.createHeaderSet();
request.createAuthenticationChallenge(
  "Test Password", true, false);
// Set the Authenticator to respond to the
// Authentication Response header
conn.setAuthenticator(new ClientAuthenticator());
// Connect to the server
HeaderSet reply = conn.connect(request);
// Verify that the server accepted the connection
if (reply.getResponseCode() !=
  ResponseCodes.OBEX_HTTP_OK) {
  conn.close();
  throw new IOException("Connection Failed (" +
    Integer.toHexString(reply.getResponseCode())+")");
}

return conn;
}
```

To respond to an authentication challenge, an OBEX server must specify the `Authenticator` in the call to `acceptAndOpen()`. The code below sets the `Authenticator` to handle the authentication headers in this way. (The `ServerAuthenticator` class is defined later in this chapter.)

```
SessionNotifier waitForConnection(String connString)
  throws IOException {
  // Establish the server connection object
  SessionNotifier notifier = (SessionNotifier)
    Connector.open(connString);
```

```
  // Wait for the client to connect
  notifier.acceptAndOpen(new RequestHandler(), new
    ServerAuthenticator());
  return notifier;
}
```

When the server receives a CONNECT request with an
AUTHENTICATION_CHALLENGE header, the onAuthentication-
Challenge() method is called on the Authenticator object specified in acceptAndOpen(). The onAuthenticationChallenge()
method is written as part of an implementation of the Authenticator
interface and must return the user name and password to the implementation via a PasswordAuthentication object. The code below
shows this process.

```
public class ServerAuthenticator implements Authenticator {
  public ServerAuthenticator() {
  }
  /**
   * When an AUTHENTICATION_CHALLENGE header is received,
   * pass the user name and password back to the
   * implementation.
   *
   * @param description specifies which password to use
   * @param isUserIDRequired true if the user name is
   * required; false if the user name is not required
   * @param isFullAccess true if full access will be
   * granted; false if full access will not be granted
   * @return the user name and password or null if the
   * description does not specify "Test Password"
   */
  public PasswordAuthentication onAuthenticationChallenge(
    String description, boolean isUserIDRequired,
    boolean isFullAccess) {
    if (description.equals("Test Password")) {
      return new PasswordAuthentication(
```

```
        new String("Bob").getBytes(),
        new String("GoodPassword").getBytes());
  }

  return null;
  }

  public byte[] onAuthenticationResponse(byte[] username) {
    return null;
  }
}
```

After calling onAuthenticationChallenge, the OBEX API implementation on the server then invokes the onConnect() method. This allows the server to include additional headers in the reply or reject the connection.

```
public class RequestHandler extends
  ServerRequestHandler {
  public RequestHandler() {
  }

  /**
   * Accept the connection. This method is called each
   * time a CONNECT  request is received.
   *
   * @param request ignored
   * @param reply set the TYPE header
   * @return always return OBEX_HTTP_OK
   */
  public int onConnect(HeaderSet request, HeaderSet
    reply) {
    reply.setHeader(HeaderSet.TYPE, "text/text");
    return ResponseCodes.OBEX_HTTP_OK;
  }
}
```

After the server sends the response in an AUTHENTICATION_ RESPONSE header, the client's OBEX API implementation invokes the onAuthenticationResponse() method on the client's

Authenticator object. The onAuthenticationResponse() method allows the client to pass the OBEX API implementation the correct password. After the OBEX API implementation receives a non-null password, the implementation validates the password. If the password is valid, the connect() method returns the HeaderSet received from the server. If the onAuthenticationResponse() method returns null, the authentication fails. If the authentication fails because null is returned or the wrong password was supplied, the call to connect() throws an IOException specifying that the authentication failed. The following code processes an AUTHENTICATION_RESPONSE header.

```
public class ClientAuthenticator implements
  Authenticator {
  public ClientAuthenticator() {
  }
  /**
   * Validates the password by returning the valid
   * password.
   * @param username the user name provided; null if no
   * user name was included in the header
   * @return the password for the user name specified;
   * null if the user name or password is not valid
   */
  public byte[] onAuthenticationResponse(byte[]
    username) {
    // Checks to see if the only valid user name was
    // provided, otherwise fail the authentication
    // request by returning null
    if ((username == null) || (!new
      String(username).equals("Bob"))) {
      return null;
    }
    return new String("GoodPassword").getBytes();
  }
```

```
public PasswordAuthentication onAuthenticationChallenge(
   String description, boolean isUserIDRequired,
   boolean isFullAccess) {
   return null;
}
}
```

When a server wants to authenticate a client, the preceding process is followed, with two exceptions. First, an onXXX() method, such as onConnect() or onPut(), must add the AUTHENTICATION_ CHALLENGE header by calling createAuthenticationChallenge() and return the OBEX_HTTP_UNAUTHORIZED response code. Second, the OBEX API implementation invokes the onAuthenticationFailure() method of the ServerRequestHandler specified instead of throwing an IOException as the client implementation does if the authentication fails. (Here's a tip: If a server uses OBEX authentication, it is easier to detect authentication failures if a new ServerRequestHandler object is passed to acceptAndOpen() for each connection.)

5.4 Summary

OBEX was defined by IrDA and adopted by the Bluetooth SIG. The GOEP defines OBEX within the Bluetooth world. Many different profiles have been defined on top of GOEP. For Bluetooth devices, OBEX uses the RFCOMM protocol as the transport protocol.

OBEX is built on a request and response scheme. The client drives the connection by issuing requests to the server. The server can accept or reject the request using any of the HTTP response codes. CONNECT, PUT, GET, SETPATH, DELETE, CREATE-EMPTY, and DISCONNECT are the valid operations that may be performed by a client. A client session begins with a CONNECT request and ends with a DISCONNECT request. Between the CONNECT and DISCONNECT request, any number of PUT, GET, SETPATH, DELETE, and CREATE-EMPTY operations can occur. All data is sent within OBEX headers. Any of these headers can be included in any of the operations.

OBEX provides a structured way to send data between embedded devices. Although RFCOMM works by sending bytes between devices via a stream, OBEX sends logical objects, not Java objects, between devices. By using OBEX, a developer does not need to worry about adding structure to a stream of bits. The developer needs to worry only about which headers to send and retrieve.

The OBEX API defined within JABWT is a separate API from the Bluetooth API. For this reason, the OBEX API was designed to be transport neutral. The OBEX API is built on the GCF defined by CLDC. The connection string passed to `Connector.open()` specifies which transport to use. For client connections, a `ClientSession` object is returned. A `SessionNotifier` object is returned for server connections.

With the `ClientSession` object returned by `Connector.open()`, `HeaderSet` objects can be created to send headers in any request. After `Connector.open()` is called, only the transport layer connection has been established. To establish an OBEX session, the `connect()` method is called. If the server accepts the connection request, the `put()`, `get()`, `setPath()`, or `delete()` method can be called to issue the associated request to the server. After the client finishes communicating with the server, the `disconnect()` method should be called to end the OBEX session and should be followed by the `close()` method to close the transport layer connection.

OBEX server connections work slightly differently. To process requests from a client, the server application must provide a class that extends the `ServerRequestHandler` class to the `SessionNotifier`'s `acceptAndOpen()` method. Requests from the client are passed to the OBEX server via events to the `ServerRequestHandler` class. The `onConnect()`, `onPut()`, `onGet()`, `onSetPath()`, `onDelete()`, and `onDisconnect()` methods are called when the associated request is received from the client. Within the `onXXX()` method, the server can set any headers to send in the reply along with the response code.

OBEX provides a mechanism for authentication. This method is different from Bluetooth authentication. OBEX authentication uses a challenge and response scheme. Within the OBEX API, authentication starts with the `Authenticator` interface. When an authentication challenge or response is received, the appropriate method is called in the `Authenticator` object specified to the JABWT implementation. The JABWT implementation handles the details of packaging the response and determining whether the response was correct.

6 CHAPTER

Device Discovery

This chapter covers the following topics:

- Retrieving information about the local device
- Why is device discovery needed?
- Making a device discoverable
- Retrieving devices without an inquiry
- How to start an inquiry
- Changing security on a link
- Working with remote devices

6.1 Overview

Because the typical Bluetooth radio is part of a mobile device, a Bluetooth device must be able to dynamically locate other nearby Bluetooth devices. A Bluetooth device must also be able to determine what services are on the devices found. The Bluetooth specification separates discovery of devices and discovery of services into separate processes. In the device discovery process, the local Bluetooth device finds the other Bluetooth devices in the area. In the service discovery process, the Bluetooth device determines which services the other devices have running on them.

In Bluetooth terms, device discovery is known as an *inquiry*. When a Bluetooth device issues an inquiry request, the other devices in the area respond to the inquiry request depending on their discoverable mode. These devices respond with their Bluetooth address and class of device record. The Bluetooth address is a 6-byte unique identifier assigned to every Bluetooth device by the manufacturer. The class of device record describes the type of Bluetooth device and provides a general indication of the types of services available on the device.

The Bluetooth SIG has defined two types of inquiries: general and limited. A general inquiry is used to find all the Bluetooth devices in an area. A limited inquiry is used to find all devices in an area that are discoverable for only a limited length of time. A general inquiry is similar to asking all people in a room to say their names. A limited inquiry is similar to asking all people in a room to say their names only if they are accountants. Which devices respond to an inquiry request depends on the discoverable mode of the device. A Bluetooth

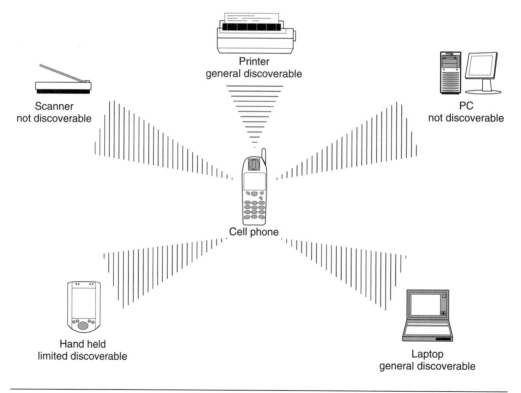

Figure 6.1   Cell phone performs a general inquiry.

device can be general, limited, or not discoverable. A general discoverable device responds only to general inquiries. Limited discoverable devices respond to general and limited inquiries. A device cannot respond to any inquiries if it is not discoverable.

During a general inquiry, the device performing the inquiry asks all the devices in the area to respond to a general inquiry request. Only devices that are limited or general discoverable respond to the request. For example, in Figure 6.1, the cell phone issues the general inquiry request. Although they may receive the request, the PC and scanner do not respond to the request because they are in the not discoverable mode (Figure 6.2). The remaining devices in the area do respond to the cell phone with their Bluetooth addresses and class of device record.

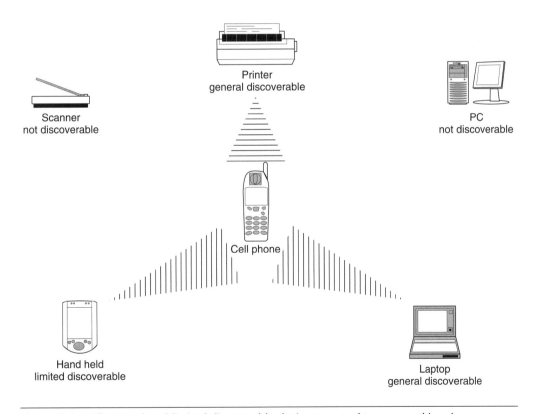

Figure 6.2   All general and limited discoverable devices respond to a general inquiry.

6.2 API Capabilities

JABWT provides two approaches to device and service discovery. First, JABWT provides methods that allow full control over device and service discovery. The second approach leaves the device and service discovery up to the JABWT implementation. This separation was created to allow developers to develop Bluetooth profiles with the API and to optimize their applications while also allowing developers to quickly get a Bluetooth application up and running. (The second approach is presented in Section 7.3.8.)

Device discovery starts with the JABWT application specifying the type of inquiry to perform, either general or limited. (Additional inquiry types are possible but have yet to be defined by the Bluetooth SIG.) The API returns each device found during an inquiry back to the application as the device is found via a `deviceDiscovered()` event. The device found and the class of device record of the remote device are returned.

The class of device record specifies the type of device responding and the services available on the device. The class of device record consists of the major service class, major device class, and minor device class. The major service class defines the services available on a device. The following list defines all the major service classes defined by the Bluetooth SIG when this book was published.

- Positioning
- Networking
- Capturing
- Object Transfer
- Audio
- Information
- Limited Discoverable
- Rendering
- Telephony

A device can have multiple major *service* classes. In other words, a device can have an audio and a telephony service running on it at the same time, which makes the major service classes audio and telephony. The major device class and minor device class describe the physical type of device of which the Bluetooth radio is a part. A device can have only a single major *device* class. The Bluetooth SIG has defined the following major device classes:

- Computer (e.g., desktop, notebook, PDA, organizers)
- Phone (e.g., cellular, cordless, pay phone, modem)
- LAN/network access point
- Audio/video (e.g., headset, speaker, stereo, video display)
- Peripheral (e.g., mouse, joystick, keyboard)
- Imaging (e.g., printer, scanner, camera, display)
- Miscellaneous
- Wearable
- Toy
- Medical

The minor device class is defined on the basis of the major device class. The list of all the minor device classes is available from the Bluetooth SIG [34]. The minor device class is simply a more specific description of the device. For example, a device can be classified as an imaging device. The minor device class specifies whether the device is a camera, scanner, or printer. This system allows developers to perform service searches on only the devices that fit their needs. In the example, the developer is able to eliminate the camera and scanner devices from the service search if the developer is looking for a printing service.

JABWT provides a simpler approach to retrieving remote devices that eliminates the need for performing a complete inquiry. When it performs an inquiry, an application must wait 8 to 10 seconds for a 95% chance of finding all the devices in the area. In addition to taking time, an inquiry is power consuming, which is a major concern for embedded, battery-powered devices. JABWT allows an application to retrieve a list of devices that would "likely" be in an area without performing an inquiry. These devices are called *predefined*. There are two types of predefined devices: pre-known and cached. Pre-known devices are devices with which the local device frequently interacts. Pre-known devices are set in the BCC. (Section 3.2 provides a full description of the BCC.) For example, a PDA frequently synchronizes with a desktop computer. The user of the PDA would specify the desktop computer as a pre-known device. A pre-known device does not guarantee that the device is currently reachable or available, but it does give an educated guess. The second type of device that can be retrieved without an inquiry is a cached device. Cached devices are Bluetooth devices found with a previous inquiry. The inquiry does not have to be performed by the current application. The cached devices might have been found by a previous inquiry performed by another application on this same Bluetooth device.

6.3 Programming with the API

The LocalDevice class provides access to the local Bluetooth device. There is only a single LocalDevice object for the entire JABWT implementation. This LocalDevice object provides methods of retrieving information about the local device and a gateway for starting the different discovery processes.

6.3.1 Accessing the Local Device

Retrieving information and manipulating a local Bluetooth device begin with retrieving the LocalDevice object. The LocalDevice class has a private constructor, which prevents an application from creating a new LocalDevice object on its own. For an application to retrieve the LocalDevice object for the JABWT implementation, the application calls the LocalDevice.getLocalDevice() method. The getLocal-Device() method may throw a BluetoothStateException if the Bluetooth stack or radio is not working properly.

Once the LocalDevice has been retrieved, more information can be gathered about the local device, such as the device's Bluetooth address, friendly name, current discoverable mode, and class of device record. The getBluetoothAddress() method returns the device's address. The getFriendlyName() method returns the device's user-friendly name or returns null if the name could not be retrieved. The get-Discoverable() method returns the local device's current discoverable mode. The DiscoveryAgent class contains constants for general discoverable (GIAC, which stands for General Inquiry Access Code), limited discoverable (LIAC, which stands for Limited Inquiry Access Code), and not discoverable (NOT_DISCOVERABLE). Other discoverable modes are possible but have not been defined by the Bluetooth SIG. Finally, the class of device record can be retrieved by means of the getDeviceClass() method. The getDeviceClass() method returns null if the class of device record could not be retrieved. (Section 6.3.4 describes manipulating the class of device record through a DeviceClass object.)

The LocalDevice class has a method for requesting a different discoverable mode. The setDiscoverable() method takes the requested discoverable mode as an argument. The discoverable mode can be DiscoveryAgent.GIAC, DiscoveryAgent.LIAC, Discovery-Agent.NOT_DISCOVERABLE, or any discoverable mode in the range 0x9E8B00 to 0x9E8B3F. The setDiscoverable() method returns true if the discoverable mode changes to the requested mode. This method returns false if the BCC denies the request or the local device does not support the requested discoverable mode. The set-Discoverable() method throws a BluetoothStateException if the requested change cannot occur at this time because the device is in a state that does not allow the change.

What happens when one device asks for limited discoverable while another asks for general discoverable? This is left up to the BCC. JABWT does not place requirements on how the BCC resolves conflicting discoverable requests. The possible approaches include but are not limited to the following:

1. The first requested discoverable mode is honored until the application ends.
2. The last request is always honored.
3. One discoverable mode has a higher priority than another. For example, general discoverable has a higher priority than limited discoverable. Not discoverable is the lowest priority. A request to a higher priority is honored, whereas a request for a lower priority is denied.

The LocalDevice class also has the getProperty() method for retrieving additional information on the capabilities of the Bluetooth radio and stack. The `LocalDevice.getProperty()` method works similarly to the `System.getProperty()` method. The `LocalDevice.getProperty()` method takes, as a `String`, the specific property whose value is to be retrieved. The `getProperty()` method returns the value of the property as a `String`. Table 6.1 lists all the properties available through the `getProperty()` method. A JABWT implementation can add parameters to the `getProperty()` method but must support at a minimum all the properties in Table 6.1. The argument is case sensitive.

To show how to use these methods within an application, the `BluetoothInfoMIDlet` follows. The `BluetoothInfoMIDlet` displays a `Form` with the Bluetooth address, friendly name, current discoverable mode, and all the properties of the local Bluetooth device. The first step to creating the MIDlet is creating the `BluetoothInfoMIDlet` class and adding code to its `startApp()` method to display a `Form`. The next step is to add an exit `Command` to the `Form` to destroy the MIDlet.

```java
package com.jabwt.book;
import java.lang.*;
import javax.microedition.midlet.*;
import javax.microedition.lcdui.*;
import javax.bluetooth.*;

public class BluetoothInfoMIDlet extends BluetoothMIDlet {
  /**
   * Called when the MIDlet is started to display the
   * properties of the MIDlet on a Form.
   */
  public void startApp()
   throws MIDletStateChangeException {
    Display currentDisplay = Display.getDisplay(this);
    Form infoForm = new Form("Device Info");
    currentDisplay.setCurrent(infoForm);

    // Add the exit command and set the listener
    infoForm.addCommand(new Command("Exit",
      Command.EXIT, 1));
    infoForm.setCommandListener(this);
  }
}
```

Table 6.1 Properties Available through `LocalDevice.getProperty()`

Property	Description	Valid Value
bluetooth.api.version	The version of JABWT that is supported. This property does not relate to the Bluetooth specification number.	Depends on the JABWT version
bluetooth.master.switch	Is master/slave switch allowed?	"true" or "false"
bluetooth.sd.attr.retrievable.max	Maximum number of service attributes to be retrieved per service record.	A base 10 integer (e.g., "1", "2")
bluetooth.connected.devices.max	Maximum number of connected devices supported.	A base 10 integer (e.g., "1", "2")
bluetooth.l2cap.receiveMTU.max	Maximum receiveMTU size in bytes supported in L2CAP	A base 10 integer (e.g., "1", "2")
bluetooth.sd.trans.max	Maximum number of concurrent service discovery transactions.	A base 10 integer (e.g., "1", "2")
bluetooth.connected.inquiry.scan	Can the local device respond to an inquiry request while the device has established a link to another device?	"true" or "false"
bluetooth.connected.page.scan	Can the local device accept a connection from a remote device if it is already connected to another remote device?	"true" or "false"
bluetooth.connected.inquiry	Can the local device start an inquiry while it is connected to another device?	"true" or "false"
bluetooth.connected.page	Can the local device establish a connection to a remote device if the local device is already connected to another device?	"true" or "false"

Now that the `BluetoothInfoMIDlet` has been created, the `getBluetoothInfo()` method is added. The `getBluetoothInfo()` method retrieves the Bluetooth device information and displays it on the screen. This method first retrieves the `LocalDevice` object. Next, the Bluetooth address is retrieved and displayed on the screen. Before the user-friendly name returned from `getFriendlyName()` is displayed, the `getBluetoothInfo()` method must verify that `null` was not returned. If `null` was returned, the appropriate message is added to the `Form`; otherwise, the user-friendly name is displayed. A `switch` statement is used to resolve the value of the discoverable mode with its associated name. Finally, each of the properties available via the `LocalDevice.getProperty()` method is retrieved and appended to the `Form`. The values returned from `getProperty()` do not need to be checked for `null` because these values are required to be part of the JABWT implementation.

```
public class BluetoothInfoMIDlet extends BluetoothMIDlet {
  public void startApp()
    throws MIDletStateChangeException {
    Display currentDisplay = Display.getDisplay(this);
    Form infoForm = new Form("Device Info");
    currentDisplay.setCurrent(infoForm);


    getBluetoothInfo(infoForm);


    // Add the exit command and set the listener
    infoForm.addCommand(new Command("Exit",
      Command.EXIT, 1));
    infoForm.setCommandListener(this);
  }


  /**
   * Displays the Bluetooth device information on the screen.
   * @param f display the information on Form f
   */
  private void getBluetoothInfo(Form f) {
    LocalDevice local = null;
    // Retrieve the local Bluetooth device object
    try {
```

```
    local = LocalDevice.getLocalDevice();
} catch (BluetoothStateException e) {
  f.append("Failed to retrieve the local device (" +
    e.getMessage() + ")");
  return;
}
// Retrieve the Bluetooth address
f.append("Address: " + local.getBluetoothAddress());
f.append("\n");
// Retrieve the Bluetooth friendly name
String name = local.getFriendlyName();
if (name == null) {
  f.append("Name: Failed to Retrieve");
} else {
  f.append("Name: " + name);
}
f.append("\n");
// Retrieve the current discoverable mode
int mode = local.getDiscoverable();
StringBuffer text = new StringBuffer(
  "Discoverable Mode: ");
switch (mode) {
case DiscoveryAgent.NOT_DISCOVERABLE:
  text.append("Not Discoverable");
  break;
case DiscoveryAgent.GIAC:
  text.append("General");
  break;
case DiscoveryAgent.LIAC:
  text.append("Limited");
  break;
default:
  text.append("0x");
  text.append(Integer.toString(mode, 16));
  break;
}
```

```
f.append(text.toString());
f.append("\n");

// Retrieve all the default properties
// and display them on the screen.
f.append("API Version: " +
  local.getProperty("bluetooth.api.version"));
f.append("\n");
f.append("Master Switch Supported: " +
  local.getProperty("bluetooth.master.switch"));
f.append("\n");
f.append("Max Attributes: " +
  local.getProperty(
  "bluetooth.sd.attr.retrievable.max"));
f.append("\n");
f.append("Max Connected Devices: " +
  local.getProperty(
  "bluetooth.connected.devices.max"));
f.append("\n");
f.append("Max Receive MTU: " +
  local.getProperty(
  "bluetooth.l2cap.receiveMTU.max"));
f.append("\n");
f.append("Max Service Discovery Transactions: " +
  local.getProperty("bluetooth.sd.trans.max"));
f.append("\n");
f.append("Connection Options\n");
f.append(" Inquiry Scan Supported: " +
  local.getProperty(
  "bluetooth.connected.inquiry.scan"));
f.append("\n");
f.append(" Page Scan Supported: " +
  local.getProperty(
  "bluetooth.connected.page.scan"));
f.append("\n");
f.append(" Inquiry Supported: " +
  local.getProperty("bluetooth.connected.inquiry"));
```

```
      f.append("\n");
      f.append(" Page Supported: " +
        local.getProperty("bluetooth.connected.page"));
      f.append("\n");
   }

}
```

The `BluetoothInfoMIDlet` displays a great deal of information about the local Bluetooth device and its JABWT implementation. Figure 6.3 shows a screen shot of the `BluetoothInfoMIDlet` running.

Simple Device Discovery

JABWT provides device and service discovery capabilities via the `DiscoveryAgent` class and the `DiscoveryListener` interface. Each device has a single `DiscoveryAgent` object. The `DiscoveryAgent`

Figure 6.3   `BluetoothInfoMIDlet` running in Motorola simulator (emulation only).

provides methods to start device and service searches. The `Discovery-Listener` interface is used by the `DiscoveryAgent` to pass devices and services back to the application as they are found. This `Discovery-Agent` object is retrieved from the local device via the `LocalDevice.getDiscoveryAgent()` method. The following method shows how to retrieve the `DiscoveryAgent` object for a Bluetooth device.

```
public static DiscoveryAgent getLocalDiscoveryAgent() {
  try {
    /*
     * Retrieve the local Bluetooth device object for
     * this device. This method may throw a
     * BluetoothStateException if the local device
     * could not be initialized.
     */
    LocalDevice local = LocalDevice.getLocalDevice();
    DiscoveryAgent agent = local.getDiscoveryAgent();

    return agent;
  } catch (BluetoothStateException e) {
    return null;
  }
}
```

The `getLocalDiscoveryAgent()` method must catch a `Bluetooth-StateException` because `LocalDevice.getLocalDevice()` may throw it. Once the `LocalDevice` is retrieved, the call to `getDiscovery-Agent()` returns the `DiscoveryAgent` associated with the `LocalDevice`. Multiple calls to `getDiscoveryAgent()` return the same object.

The `DiscoveryListener` interface is the other intricate part of device and service discovery in JABWT. The `DiscoveryListener` interface is implemented by an application to receive devices and service records as they are discovered. The interface also provides an application with a notification when the inquiry or service search has completed.

The easiest way to retrieve a list of `RemoteDevices` is using the `retrieveDevices()` method in the `DiscoveryAgent` class. The `retrieveDevices()` method returns the list of pre-known devices if the `DiscoveryAgent.PREKNOWN` argument is used and the list of cached devices if the `DiscoveryAgent.CACHED` option is set. Pre-known devices are devices with which the local device commonly interacts. Devices can be registered as pre-known in the BCC. There is no guarantee that a pre-known device is in the area or can be connected to. Cached devices are devices that have been found via a previous inquiry. How many devices are cached and for how long is implementation dependent. Because of these facts, there is no guarantee that a cached or pre-known device is currently available, but the retrieveDevices() method is a quick way to get to the service search phase.

To show how to use the retrieveDevices() method, the DiscoveryMIDlet is created to display all the pre-known and cached devices. To add this capability, the startApp() method creates a List and makes it the current displayable. The startApp() method calls the addDevices() method to retrieve the pre-known and cached devices and to print out the Bluetooth addresses of each of these devices. To identify which devices are pre-known and which are cached, "-P" or "-C" is appended to each device's Bluetooth address to represent pre-known and cached, respectively. Figure 6.4 shows the DiscoveryMIDlet running in simulation. (Note: Pre-known devices may be set by editing the device in the Impronto Simulator and pressing the "Pre-Known" button.)

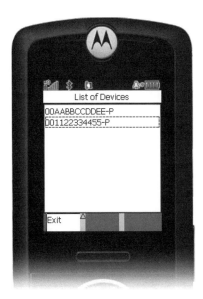

Figure 6.4   DiscoveryMIDlet retrieving list of pre-known and cached devices (emulation only).

```java
package com.jabwt.book;

import java.lang.*;
import java.util.*;
import java.io.*;
import javax.microedition.midlet.*;
import javax.microedition.lcdui.*;
import javax.bluetooth.*;

public class DiscoveryMIDlet extends BluetoothMIDlet {
  /**
   *The List of remote devices
   */
  private List deviceList;
  /**
   * The DiscoveryAgent for the local device.
   */
  private DiscoveryAgent agent;
  /**
   * Retrieves the list of pre-known and cached devices.
   * Updates the display to show the list of devices.
   */
  public void startApp()
   throws MIDletStateChangeException {
    // Create a new List and set it to the current
    // displayable
    deviceList = new List("List of Devices", List.IMPLICIT);
    deviceList.addCommand(new Command("Exit",
      Command.EXIT, 1));
    deviceList.setCommandListener(this);
    Display.getDisplay(this).setCurrent(deviceList);

    // Retrieve the DiscoveryAgent object. If
    // retrieving the local device causes a
    // BluetoothStateException, something is wrong so
    // stop the app from running.
    try {
      LocalDevice local = LocalDevice.getLocalDevice();
      agent = local.getDiscoveryAgent();
```

```
    } catch (BluetoothStateException e) {
      // Prevent the application from starting if the
      // Bluetooth device could not be retrieved.
      throw new MIDletStateChangeException(
        "Unable to retrieve local Bluetooth device.");
    }
    addDevices();
  }
  /**
   * Updates the List of devices with the cached and
   * pre-known devices.
   */
  private void addDevices() {
    // Retrieve the pre-known devices array and append
    // the addresses of the Bluetooth devices. If there
    // are no pre-known devices, move on to cached
    // devices.
    RemoteDevice[] list = agent.retrieveDevices(
      DiscoveryAgent.PREKNOWN);

    if (list != null) {
      for (int i = 0; i <list.length; i++) {
        String address = list[i].getBluetoothAddress();
        deviceList.insert(0, address + "-P", null);
      }
    }
    // Retrieve the cached device array and add the
    // addresses to the list.
    list = agent.retrieveDevices(DiscoveryAgent.CACHED);
    if (list != null) {
      for (int i = 0; i < list.length; i++) {
        String address = list[i].getBluetoothAddress();
        deviceList.insert(0, address + "-C", null);
      }
    }
  }
}
```

6.3.2   Device Discovery Via Inquiry

Starting an inquiry is more complicated than simply retrieving a list of devices. An inquiry requires the Bluetooth radio to issue requests for all devices in the area to respond according to their discoverable mode. This process uses power and can prevent the radio from being used for other purposes. Before an application can request an inquiry, the application must implement the `DiscoveryListener` interface. This interface requires that the `deviceDiscovered()` and the `inquiryCompleted()` methods be implemented for device discovery. To actually start an inquiry, the `startInquiry()` method is used. The `startInquiry()` method takes, as arguments, the type of inquiry and an implementation of the `DiscoveryListener` interface.

JABWT provides constants for the two types of inquiries defined by the Bluetooth SIG. The `DiscoveryAgent.GIAC` inquiry type should be passed to `startInquiry()` for a general inquiry. A general inquiry locates all devices in the general or limited discoverable mode. For a limited inquiry, the `DiscoveryAgent.LIAC` inquiry type should be used. Again, a limited inquiry finds all devices in the area in the limited discoverable mode. A device is placed in the limited discoverable mode when it is discoverable for only a short time. In addition to general and limited inquiry, the Bluetooth SIG specified that other inquiry access codes may be defined in the range 0x9E8B00 to 0x9E8B3F in the future. Therefore, `startInquiry()` accepts any value in this range.

The `startInquiry()` method returns `true` if the inquiry is started. Because not all devices support all the inquiry access codes, the `startInquiry()` method returns `false` if a valid inquiry access code is provided but the code is not supported. Valid inquiry access codes are in the range 0x9E8B00 to 0x9E8B3F along with GIAC and LIAC. The `startInquiry()` method throws an `IllegalArgumentException` if an invalid inquiry access code is passed as an argument. The `start-Inquiry()` method throws a `BluetoothStateException` if the inquiry could not be started because the device is in a state that does not allow an inquiry to be completed. A device may be in such a state when the device is already performing an inquiry or when the local device is already connected to another device.

After starting the inquiry using `startInquiry()`, JABWT returns devices to the application via `deviceDiscovered()` events. A `deviceDiscovered()` event occurs every time a remote Bluetooth device is found by the Bluetooth radio. The `deviceDiscovered()` event provides the `RemoteDevice` object and associated `DeviceClass` object each time the event occurs. The `RemoteDevice` object provides the Bluetooth address of a remote Bluetooth device along with methods to retrieve the friendly name and security controls for the remote device. The `DeviceClass` object contains the class of device of the `RemoteDevice`. (The `DeviceClass` object is explained further in Section 6.3.4.)

JABWT provides a way to cancel an inquiry. An application may want to cancel an inquiry once it finds a specific Bluetooth device or if the application is paused or destroyed. The `cancelInquiry()` method cancels an inquiry. To prevent one application from canceling the inquiry of another Bluetooth application, the `cancelInquiry()` method takes one parameter, the `DiscoveryListener` object used when the inquiry was started. The `cancelInquiry()` method returns `true` if the inquiry was canceled. If `cancelInquiry()` returns `false`, an inquiry could not be found associated with the `DiscoveryListener` provided, so no inquiry is canceled.

To notify the application that the inquiry has been completed, the `inquiryCompleted()` event was added to JABWT. The `inquiryCompleted()` event provides the reason the inquiry ended as an argument to the method. The `DiscoveryListener.INQUIRY_COMPLETED` reason is specified if the inquiry completes normally. The `DiscoveryListener.INQUIRY_TERMINATED` reason is passed as part of the `inquiryCompleted()` event if the inquiry was canceled by the application using the `cancelInquiry()` method. The call to `cancelInquiry()` is a non-blocking call. The `inquiryCompleted()` event occurs independently of the `cancelInquiry()` method ending. Finally, the `inquiryCompleted()` event receives a `DiscoveryListener.INQUIRY_ERROR` reason if an error occurs during processing of the inquiry.

The following code shows a simple MIDlet that starts an inquiry and displays all the devices that respond to the inquiry request. The Bluetooth address of each device is displayed in a `List`. When the inquiry ends and the `inquiryCompleted()` method is called, an

`Alert` appears to notify the user that the inquiry has ended. If an error occurs during processing of the MIDlet, an `Alert` is displayed to the user to notify the user of the error.

```
public class DiscoveryMIDlet extends BluetoothMIDlet
  implements DiscoveryListener {
```

```
  /**
   * Retrieves the list of pre-known and cached devices.
   * Updates the display to show the list of devices.
   */
  public void startApp()
   throws MIDletStateChangeException {
    // Create a new List and set it to the
    // current displayable
    deviceList = new List("List of Devices",
      List.IMPLICIT);
    deviceList.addCommand(new Command("Exit",
      Command.EXIT, 1));
    deviceList.setCommandListener(this);
    Display.getDisplay(this).setCurrent(deviceList);

    // Retrieve the DiscoveryAgent object. If
    // retrieving the local device causes a
    // BluetoothStateException, something is wrong
    // so stop the app from running.
    try {
      LocalDevice local = LocalDevice.getLocalDevice();
      agent = local.getDiscoveryAgent();
    } catch (BluetoothStateException e) {
      // Prevent the application from starting if
      // the Bluetooth device could not be retrieved.
      throw new MIDletStateChangeException(
        "Unable to retrieve local Bluetooth device.");
    }
```

```
addDevices();
```

```
try {
  agent.startInquiry(DiscoveryAgent.GIAC, this);
} catch (BluetoothStateException e) {
  throw new MIDletStateChangeException(
    "Unable to start the inquiry");
}
```

```
}
...
```

```
/**
 * Called each time a new device is discovered.
 * This method prints the device's Bluetooth
 * address to the screen.
 *
 * @param device the device that was found
 * @param cod the class of device record
 */
public void deviceDiscovered(RemoteDevice device,
  DeviceClass cod) {
  String address = device.getBluetoothAddress();
  deviceList.insert(0, address + "-I", null);
}
/**
 * Called when an inquiry ends. This method
 * displays an Alert to notify the user the inquiry
 * ended. The reason the inquiry ended is displayed
 * in the Alert.
 *
 * @param type the reason the inquiry completed
 */
public void inquiryCompleted(int type) {
  Alert dialog = null;
```

```
    // Determine if an error occurred. If one did
    // occur display an Alert before allowing the
    // application to exit.
    if (type != DiscoveryListener.INQUIRY_COMPLETED) {
      dialog = new Alert("Bluetooth Error",
        "The inquiry failed to complete normally",
        null, AlertType.ERROR);
    } else {
      dialog = new Alert("Inquiry Completed",
        "The inquiry completed normally", null,
        AlertType.INFO);
    }
    dialog.setTimeout(Alert.FOREVER);
    Display.getDisplay(this).setCurrent(dialog);
  }
  public void servicesDiscovered(int transID,
    ServiceRecord[] record) {
  }
  public void serviceSearchCompleted(int transID, int
    type) {
  }
```

```
}
```

Most of the `DiscoveryMIDlet` code is required by the MIDP specification. The important parts of the code are found in the `startApp()`, `device-Discovered()`, and `inquiryCompleted()` methods. The inquiry is started with `startInquiry()` in the `startApp()` method so that it occurs each time the MIDlet is made active. If started in the constructor, the inquiry occurs only when the MIDlet is created. If a `Bluetooth-StateException` occurs during retrieval of the `LocalDevice` object or the start of the inquiry, the `startApp()` method throws a `MIDlet-StateChangeException` to notify the KVM that the MIDlet is not able to run correctly. This procedure simplifies the user experience by allowing the KVM to handle the user interaction in the case of this type of error.

The `deviceDiscovered()` and `inquiryCompleted()` methods must be implemented because the `DiscoveryMIDlet` implements the `DiscoveryListener` interface. The `deviceDiscovered()` method is important because this is the method used to pass the remote devices found in the inquiry back to the MIDlet. For the purpose of this MIDlet, all this method does is get the remote device's Bluetooth address and add it to the `List`. (Figure 6.5 shows the `DiscoveryMIDlet` running.) The `inquiryCompleted()` method verifies that the inquiry completed successfully. If the inquiry did not complete properly, an `Alert` is displayed to notify the user of the error. If the inquiry did complete properly, an `Alert` saying so is displayed to the user.

Because the user may exit from the MIDlet before the inquiry ends, code must be added to cancel the inquiry. Therefore the `command-Action()` method is modified to call `cancelInquiry()`. Because calling `cancelInquiry()` when no inquiry is occurring does nothing, the `commandAction()` method calls `cancelInquiry()` every time the user exits from the MIDlet.

Figure 6.5   `DiscoveryMIDlet` after discovering devices via an inquiry (emulation only).

```
public class DiscoveryMIDlet extends BluetoothMIDlet
  implements DiscoveryListener {

...
```

```
/**
 * Called when a Command is selected. If it is an
 * Exit Command, then the MIDlet will be destroyed.
 *
 * @param c the Command that was selected
 * @param d the Displayable that was active when
 * the Command was selected
 */
public void commandAction(Command c, Displayable d) {
  if (c.getCommandType() == Command.EXIT) {
    // Try to cancel the inquiry.
    agent.cancelInquiry(this);
    notifyDestroyed();
  }
}
```

```
}
```

6.3.3   Retrieving Information from a Remote Device

A number of methods provide additional information about a remote
device. Before any of these methods can be called, a RemoteDevice
object must be created. There is no public constructor for the Remote-
Device class, so an application cannot directly instantiate a new
RemoteDevice object. The application must use one of the three
ways to get a RemoteDevice object. First, RemoteDevice objects are
created in the device discovery process. RemoteDevice objects are
passed to the application as arguments via deviceDiscovered()
events.

Second, a class that extends the RemoteDevice class can be written
and instantiated by an application. The following code does just this.

```
package com.jabwt.book;
import javax.bluetooth.*;
public class MyRemoteDevice extends RemoteDevice {
  /**
   * Creates a new RemoteDevice object based upon the
   * address provided.
   *
   * @param address the Bluetooth address
   */
  public MyRemoteDevice(String address) {
    super(address);
  }
}
```

The address provided in the constructor must be 12 hex characters with no preceding "0x." If the address is the same as that of the local device or if it contains non-hex characters, the constructor throws an Illegal-ArgumentException. If the address string is null, the constructor throws a NullPointerException. After instantiating a new MyRemoteDevice object, an application can call any of the RemoteDevice methods.

The third and final way to get a RemoteDevice object is using the RemoteDevice.getRemoteDevice() static method. The getRemote-Device() method takes, as an argument, a Bluetooth connection to a remote device. The getRemoteDevice() method returns a Remote-Device object representing the device to which the Bluetooth connection is connected. The getRemoteDevice() method throws an IOException if the connection is closed. The method throws an IllegalArgument-Exception if the connection is not a Bluetooth connection object or if the connection object is a notifier object.

After the RemoteDevice object is retrieved, the getBluetooth-Address(), isTrustedDevice(), and getFriendlyName() methods can be invoked. The getBluetoothAddress() method returns the Bluetooth address of the remote device. The getFriendlyName() method in the RemoteDevice class is different from the getFriendly-Name() of the LocalDevice method. The getFriendlyName() method of the RemoteDevice class takes a boolean argument that specifies whether the JABWT implementation should always retrieve the friendly name from the remote device or if it should retrieve the name only if the friendly name for the remote device is not known. Retrieving the friendly name requires the local device to establish a link to the remote device to retrieve the name. Because the friendly name on a device rarely changes, using a cached value if one exists eliminates the need to establish the link to the remote device. The `getFriendly-Name()` method throws an `IOException` if the remote device could not be contacted to retrieve the friendly name.

Bluetooth security can be specified after a connection is established by means of the `RemoteDevice` class. The `RemoteDevice` class provides methods for authenticating, encrypting, and authorizing a connection after a connection has been established to a remote device by the `authenticate()`, `encrypt()`, and `authorize()` methods, respectively.

The `authenticate()` method authenticates the remote device represented by the `RemoteDevice` object. The `authenticate()` method requires an existing connection to the `RemoteDevice`. If no connection exists, the `authenticate()` method throws an `IOException`. Calling `authenticate()` can cause a pairing to occur if the remote and local devices have not paired previously. The `authenticate()` method returns `true` if the remote device is authenticated; otherwise, it returns `false`. If the remote device has already been authenticated, the `authenticate()` method returns immediately with the value `true`. In other words, once a link has been authenticated, JABWT does not try to authenticate the remote device again until the link is destroyed and a new one is created.

The `encrypt()` method works slightly differently. It allows encryption on a connection to be turned on and off. This method takes two parameters: the connection to change the encryption on and a boolean specifying whether encryption should be turned on or off. The `Connection` object passed to encrypt must be a connection to the same remote Bluetooth device the `RemoteDevice` object is representing. The `Connection` object must also be a RFCOMM, L2CAP, or OBEX over RFCOMM connection. Like `authenticate()`, the `encrypt()` method returns `true` if the change succeeds and `false` if it fails.

Changing the encryption on a link is more complicated than simply authenticating a link. The request to turn on encryption can fail for a variety of reasons. First, encryption requires the link to be authenticated. If the authentication or pairing fails, then the request to turn on encryption also fails. Second, the remote device may not support or may not want encryption enabled on the link. Third, the BCC may not allow encryption on the link.

Turning off encryption is even more complicated because turning off encryption actually makes the link less secure. A request to turn off encryption may fail for two reasons. First, the remote device may require that encryption be enabled on the link. Second, the BCC may not allow encryption to be turned off. This may be a system-wide policy, or another application may be running on the device that requires the link to be encrypted. For all of these reasons, a call to `encrypt()` should be considered a request and its return value checked.

The final method in the `RemoteDevice` class that allows a change in the security of a connection is the `authorize()` method. Recall that authorization is done on a connection basis as opposed to a link basis. The `authorize()` method takes the `Connection` object to authorize. Because authorization requires a link to be authenticated, a call to `authorize()` can cause authentication and pairing if these events have not occurred on the link. After it has been verified that the link has been authenticated, the `authorize()` method requests the BCC to authorize the connection. The `authorize()` method returns `true` if the connection is authorized and `false` if the connection is not authorized.

The `RemoteDevice` class has three methods that allow an application to determine the security level on a connection. The `isAuthenticated()` and `isEncrypted()` methods return `true` if the link to the remote device has been authenticated or encrypted, respectively. Both methods return `false` if the requested security is not enabled or if there is no link between the two devices. The `isAuthorized()` method works slightly differently. This method takes the `Connection` object to check for authorization. The `isAuthorized()` method returns `true` if the connection has been authorized and `false` if it has not been authorized.

Finally, the `RemoteDevice` class contains a method that allows an application to determine whether a device is a trusted device. The `isTrustedDevice()` method returns `true` if the `RemoteDevice` object represents a trusted device. A trusted device is a device that always passes authorization. This condition is set and maintained by the BCC.

6.3.4   Using the `DeviceClass` Class

The `DeviceClass` is a unique object. It provides three methods to access the class of device record. The class of device record specifies the physical type of the device and the general services it provides. The `getServiceClasses()` method retrieves the list of all service classes on the device. Each time a service registers itself with a device, the type of service is specified in the class of device record. This procedure allows another device to identify whether a remote device may have a service it is looking for. For example, if an application is looking for a printing service, the application should look for a device that has a rendering service class. This eliminates the overhead of performing a service search on a device that does not have the requested service.

The `getServiceClasses()` method returns an integer representing all the major service classes available on a device. (The Bluetooth SIG in the Bluetooth Assigned Numbers [34] defines the major service classes.) Because a device can have multiple service classes, the Bluetooth SIG defines a service class by setting a bit in the class of device record. For example, bit 18 is set for the rendering service class. If a device has a rendering and an audio service, bits 18 and 21 are set. In this situation, `getServiceClasses()` returns an integer with bits 18 and 21 set or the value 2,359,296 (0x240000). To determine whether a device has a rendering service, bit 18 must be isolated. The following code provides a way of doing this.

```
boolean checkForRenderingService(DeviceClass d) {
  // The Rendering service bit is bit 18. Setting bit
  // 18 produces the number 0x40000.
  if ((d.getServiceClasses() & 0x40000) != 0) {
    return true;
  } else {
    return false;
  }
}
```

The `checkForRenderingService()` method isolates the rendering service bit by performing an AND on the service class of the device and a number with only bit 18 set. If the result of this AND is zero, then bit 18 is not set, and a rendering service is not available on the device. Table 6.2 lists the major service classes currently defined by the Bluetooth SIG, the bit number of the service class, and the integer value of setting only that bit. The Bluetooth SIG may add service classes in the future. Application developers should use the major service class cautiously because it gives only an indication of the types of services available on a device.

The major device class is different from the service classes. The major device class reports the physical type of device to which the Bluetooth radio is connected. Because a device cannot have more than one major device class, there is no need to check individual bits. The `getMajorDeviceClass()` returns the major device class value. At present, the Bluetooth SIG has defined seven major device classes. Table 6.3 lists

Table 6.2 Major Service Classes Defined by the Bluetooth SIG

Service Class	Type of Service	Bit Number	Hex Value
Limited Discoverable Mode	Device is in the limited discoverable mode	13	0x2000
Positioning	Location identification	16	0x10000
Networking	LAN, ad hoc, etc.	17	0x20000
Rendering	Printing, speaker, etc.	18	0x40000
Capturing	Scanner, microphone, etc.	19	0x80000
Object Transfer	V-inbox, v-folder, etc.	20	0x100000
Audio	Speaker, microphone, headset service, etc.	21	0x200000
Telephony	Cordless telephony, modem, headset service, etc.	22	0x400000
Information	Web server, Wireless Application Protocol (WAP) server, etc.	23	0x800000

Table 6.3   Major Device Classes Defined by the Bluetooth SIG

Major Device Class	Example	Hex Value
Computer	Desktop, notebook, PDA, organizer	0x100
Phone	Cellular, cordless, pay phone, modem	0x200
LAN/network access point		0x300
Audio/video	Headset, speaker, stereo, video display, VCR	0x400
Peripheral	Mouse, joystick, keyboard	0x500
Imaging	Printer, scanner, camera, display	0x600
Wearable		0x700
Toy		0x800
Medical		0x900
Miscellaneous	All other devices	0x000

all major device classes defined by the Bluetooth SIG at the time of publication of this book. The following code shows how simple it is to check for the major device class.

```
boolean checkForImaging(DeviceClass d) {
  if (d.getMajorDeviceClass() == 0x600) {
    return true;
  } else {
    return false;
  }
}
```

The minor device class returned via the `getMinorDeviceClass()` method must be interpreted on the basis of the major device class. Each major device class has a list of minor device classes that specify more information about the specific device. For example, the minor device class specifies whether a device with the computer major device class is a desktop, notebook, or PDA. The full listing of minor device classes for each major device class is available in the Bluetooth Assigned Numbers document [34].

6.4 Summary

Device discovery is a key part of any JABWT application. In Bluetooth terms, device discovery is known as *inquiry*. There are two types of inquiry: general and limited. Devices respond to inquiry requests according to their discoverable mode. There are three types of discoverable mode: not discoverable, limited discoverable, and general discoverable. When a device issues a general inquiry, all devices that are limited and general discoverable respond. When a device performs a limited inquiry, only devices that are limited discoverable respond.

All device and service discovery is started with the `DiscoveryAgent` class. The `DiscoveryAgent` class provides two different methods for discovering devices. The `retrieveDevices()` method allows applications to retrieve a list of devices found via a previous inquiry or a list of devices with which the local device frequently communicates. The `startInquiry()` method actually performs an inquiry. As devices are found, they are passed back to the application via `deviceDiscovered()` events. In addition to the devices, `deviceDiscovered()` events also pass back the class of device record. The class of device record contains information on the type of device and the services available on the device.

This chapter shows how to retrieve additional information on local and remote devices. The `LocalDevice` class provides methods that allow the applications to request the current discoverable mode, retrieve the friendly name of the local device, and retrieve information on the JABWT implementation. The `RemoteDevice` class provides similar methods for retrieving additional information on the remote device. The `RemoteDevice` class also provides methods for setting and retrieving different security settings on the link to the remote device.

7 CHAPTER   Service Discovery

This chapter covers the following topics:

- What is a Bluetooth service?
- What is a service record?
- How to perform a service search
- Retrieving additional service record attributes
- Using the simple device and service discovery API
- What is service registration?
- How are service records created and added to the SDDB?
- How are service records modified by server applications?
- What is a run-before-connect service?

7.1 Overview

After the devices in an area are discovered, the next step before connecting to a device is finding the services a device has running on it. Unlike device discovery, the service discovery process involves only a single pair of devices (Figure 7.1). The service discovery process requires the device searching for services to ask a device with services whether it has a service defined by a service record that has a specific set of attributes. If a remote device has a service with the attributes specified, the remote device returns the service record describing the service. The service record has multiple attributes. These attributes provide additional

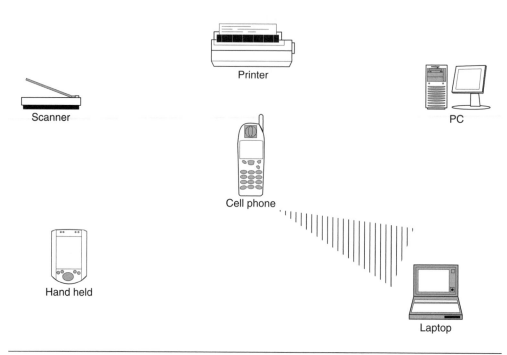

Figure 7.1   Unlike an inquiry, a service search is between only two devices.

information on a specific service. These attributes may contain anything, including information on how to connect to the service.

Service discovery follows the client-server model. A service discovery client issues a service search request to a service discovery server. The service discovery server determines whether the server has any services that meet the search criteria. For a server to know what services are available, each service registers with a service discovery database (SDDB) kept by the Bluetooth stack. When it receives a service search request, the server searches the SDDB for the specified service.

Chapter 3 lists the typical responsibilities of a Bluetooth server application:

1. Create a service record that describes the service offered by the application
2. Add the service record to the server's SDDB to make potential clients aware of this service

3. Register the Bluetooth security measures associated with a service

4. Accept connections from clients that request the service offered

5. Update the service record in the SDDB if characteristics of the service change

6. Remove or disable the service record in the SDDB when the service is no longer available

Responsibilities 1, 2, 5, and 6 compose a subset of the server responsibilities having to do with advertising a service to client devices. We call this subset *service registration*. This chapter describes the process of service registration with JABWT and the process of service discovery with JABWT. Server responsibility 3, which involves security, is discussed in Chapter 4. Server responsibility 4, which involves accepting client connections, is discussed in Chapter 4 for serial port servers, Chapter 5 for OBEX servers, and Chapter 8 for L2CAP servers.

7.1.1　Key Concepts

A service record answers the following questions:

- What kind of service does this server application offer?
- How does a client connect to this service?

Figure 3.3 is an overview of the components involved in service registration and service discovery. Service records and the Service Discovery Protocol (SDP) are described in great detail in the Bluetooth Core specification. However, the JABWT specification is the first standard API for Bluetooth service registration. The following questions about service registration were left unanswered by the Bluetooth specification:

- How are service records created?
- How are service records added to the SDDB so clients can discover them?
- How are service records modified?
- How are service records removed from the SDDB (or otherwise disabled) so clients can no longer discover them?

The Bluetooth specification did not define requirements in these areas because a standardized approach to service registration was not required for ensuring interoperability of Bluetooth devices from different manufacturers. Consequently, the mechanics of service registration were left for Bluetooth stack implementations to define. The result was a variety of different APIs for accomplishing service registration. The standard API defined by JABWT service registration makes it possible to write Bluetooth server applications that are portable across all JABWT implementations. JABWT service registration also potentially serves as a model for Bluetooth APIs in other programming languages.

7.1.2   Services Defined by Bluetooth Profiles

The Bluetooth SIG has provided profile specifications that describe standardized services. Examples of some of these services are file transfer services, business-card exchange services, and synchronization services.

If a service is defined by a Bluetooth profile, then the profile specification describes the requirements for the service record, device security, device discoverable modes, and so on. If you want to claim that your service implements a Bluetooth profile, you have to qualify your application through the Bluetooth qualification process [12].

7.1.3   Custom Services

Developers can define their own Bluetooth server applications beyond and independently of those specified in the Bluetooth profiles and make these services available to remote clients. Applications that do not claim to provide a service described in a Bluetooth profile do not need to undergo the Bluetooth qualification process. Custom services have a great deal more latitude about how they are implemented than do Bluetooth profile implementations. The developers of custom services provide the software for both communicating Bluetooth devices. The server application can be tailored to particular characteristics of the client implementation. This process is different from that for servers for Bluetooth profiles, which must be written to work with many different implementations of the client application.

7.2 API Capabilities

Once a list of devices has been retrieved via device discovery, the next step for a Bluetooth application is determining which applications or services are available on the remote Bluetooth device. JABWT provides a non-blocking way to retrieve all service records that meet a specific set of requirements on a remote Bluetooth device. A service record describes a service and is made of a set of attributes. Attributes specify how to connect to a service, the name of the service, a description of the service, and other useful information. When an application searches for a service, the application provides a set of UUIDs to search for. (A UUID is a bit sequence that uniquely identifies a characteristic of a service.) UUIDs are used to describe attributes of a service. Some UUIDs are specified by the Bluetooth specification. Other UUIDs are defined on a service-by-service basis. JABWT provides a way to specify a set of attributes to retrieve once a service is found. When a service is found that contains all the UUIDs specified, the service's service record is returned via a `servicesDiscovered()` event.

To provide a simple way to get a JABWT application up and running, JABWT defines a method that performs both device and service discovery while hiding the details of both capabilities. The `selectService()` method allows an application to specify a single UUID, which is used to locate the requested service on a remote device. The `selectService()` method returns a connection string that can be used by `Connector.open()` to connect to the service found. This is a blocking method and can take longer than ten seconds in some situations. Therefore an application should invoke this method in a separate thread to prevent the application from appearing frozen to the user.

7.2.1 Run-before-Connect Services

Ordinarily, a server application must be running and ready to accept connections before a client attempts to make a connection to the server. Server applications that have this requirement are called run-before-connect services in the JABWT specification. (The other type of services defined by JABWT, connect-anytime services, are covered in

Chapter 9.) Figure 7.2 is a Unified Modeling Language (UML) sequence diagram that illustrates the messages involved in service registration for a run-before-connect service. Each arrow in the sequence diagram is a message. The top-to-bottom ordering of the arrows indicates the time sequence of the messages. The boxes at the top of the vertical lifelines indicate the objects that send or receive the messages. (If these diagramming conventions are unfamiliar, a description of UML sequence diagrams can be found in Fowler and Scott [35].) In Figure 7.2 the boxes are all Java objects created by a JABWT program with the exception of the SDDBRecord. The SDDBRecord is a service record in the SDP server's database. The SDDBRecord is not directly accessible by a JABWT application.

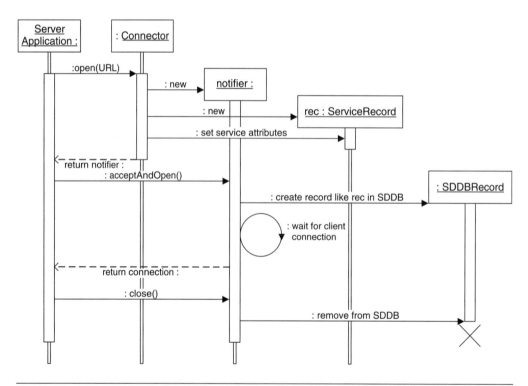

Figure 7.2   Life cycle of a service record for a run-before-connect service.

Figure 7.2 illustrates the answers that JABWT offers to the following questions about Bluetooth service records for run-before-connect services:

- How are service records created?
- How are service records added to the SDDB so clients can discover them?
- How are service records removed from the SDDB (or otherwise disabled) so clients can no longer discover them?

The methods `open()`, `acceptAndOpen()`, and `close()` shown in Figure 7.2 are part of the GCF defined by CLDC. These methods are used by a variety of Java ME applications for I/O operations. The basic approach taken by JABWT is to add additional behavior to these GCF methods so that service records are automatically created and then automatically added and removed from the SDDB, as follows:

- `Connector.open(String url)` creates a Bluetooth service record if the parameter `url` starts with

  ```
  btspp://localhost:
  btgoep://localhost:
  btl2cap://localhost:
  ```

- The first time an `acceptAndOpen()` message is sent to the notifier, a copy of the service record is added to the SDDB.
- When a `close()` message is sent to the notifier, the service record is removed from the SDDB or disabled. (The "X" at the bottom of the SDDBRecord timeline in Figure 7.2 is sequence diagram notation for deleting an object.)

One consequence of the JABWT approach is that in many cases the server application can rely only on the automatic behavior of the JABWT implementation and does not need to contain any code to explicitly manipulate service records.

7.2.2 Register a Service

This section describes the service records automatically created by the JABWT implementation for server applications. These service records allow clients to find the service and make a connection to the server. In many cases, these automatically generated service records are sufficient, and the server application does not need to take any other action. These descriptions of the default service records are provided so that developers:

- Can decide whether these service records are sufficient to advertise their services, and
- Can determine how to modify the service records when modifications are necessary

Service Records for Serial Port

Table 7.1 illustrates the service record automatically created by the JABWT implementation when a server application executes the following statement:

```
Connector.open("btspp://localhost:68EE141812D211D78EE"+
    "D00B0D03D76EC;name=SPPEx");
```

The Serial Port Profile specification contains a template for the service record used by the SPP. The JABWT implementation uses this template to create a service record and inserts the appropriate value for the RFCOMM server channel identifier into the service record. The result is a minimal but sufficient service record.

The service record in Table 7.1 shows four (attribute ID, attribute value) pairs. Each pair describes one attribute of the service. The shaded rows in Table 7.1 are the attribute IDs and the unshaded rows are the attribute values. The Bluetooth SDP uses a value between 0 and $2^{16}-1$ (65535) to represent each attribute ID in a service record, and these are shown in Table 7.1 as hexadecimal numbers. For example, the attribute ID 0x0001 indicates the ServiceClassIDList attribute ID, one of the attribute IDs defined by the Bluetooth SDP specification. Table 7.2 provides a list of the most common attribute IDs defined in the Bluetooth Assigned Numbers.

Each attribute value is a DataElement. A DataElement is a self-describing data structure that contains a type and a value. For example, the third attribute value in the service record in Table 7.1 is a simple DataElement with type String and value "SPPEx." The value "SPPEx" is extracted by the JABWT implementation from the parameter "name=SPPEx" in the connection string. The JABWT implementation uses the string "SPPEx" to construct a DataElement of type String. This DataElement is then used as the ServiceName attribute of the service record.

Table 7.1 Service Record Created by `Connector.open("btspp://localhost:...")`

ServiceClassIDList<0x0001>
DataElement(type = DATSEQ,
DataElement(type = UUID,
UUID(68EE141812D211D78EED00B0D03D76EC)
—from the connection string)
DataElement(type = UUID,
UUID(SerialPort<0x1101>)))

ProtocolDescriptorList<0x0004>
DataElement(type = DATSEQ,
DataElement(type = DATSEQ,
DataElement(type = UUID, UUID(L2CAP<0x0100>)))
DataElement(type = DATSEQ,
DataElement(type = UUID, UUID(RFCOMM<0x0003>))
DataElement(type = U_INT_1, 1
— server channel identifier. Assigned by the stack;
— added to the service record by JABWT)))

ServiceName<0x0100>
—Name of the service in the primary language of the service record
DataElement(type = STRING,
"SPPEx" —from "name=SPPEx" in the connection string)

ServiceRecordHandle<0x0000>
DataElement(type = U_INT_4,
12345 —value assigned by the SDP Server)

Table 7.2   Some Service Record Attribute IDs Defined by Bluetooth Assigned Numbers

Name	ID	Type	Description
ServiceRecordHandle	0x0000	32-bit unsigned integer	Uniquely identifies each service on a device.
ServiceClassIDList	0x0001	A DataElement sequence of UUIDs	Defines the service classes that describe the service. The service classes are defined by the Bluetooth SIG.
ServiceID	0x0003	UUID	Uniquely identifies the service instance associated with this service record.
ProtocolDescriptorList	0x0004	A DataElement sequence of DataElement sequences of UUIDs and optional parameters	Describes the protocols to use to connect to the service.
ServiceInfoTimeToLive	0x0007	32-bit unsigned integer	Defines the length of time this service record is valid and will remain unchanged.
ServiceAvailability	0x0008	32-bit unsigned integer	Describes the relative availability of the service to accept additional connections.
BluetoothProfileDescriptorList	0x0009	A DataElement sequence of DataElement sequence pairs	Specifies all the profiles this service implements.
DocumentationURL	0x000A	URL	A URL that points to the documentation for the service.
IconURL	0x000C	URL	A URL that points to an icon that may be used to represent the service.
ServiceName	0x0100	String	The name of the service in the primary language of this service record.
ServiceDescription	0x0101	String	A description of the service in the primary language of this service record.

A DataElement may be one of the following types:

- Null
- Integer (1, 2, 4, 8, and 16 byte)
- Unsigned integer (1, 2, 4, 8, and 16 byte)
- URL
- UUID
- String
- Boolean
- DATALT—DataElement alternative
- DATSEQ—DataElement sequence

Three DataElement types require further explanation: UUID, Data-Element sequence, and DataElement alternative.

A DataElement sequence is a list of DataElements in which all elements are part of the definition. In other words, a DataElement sequence is an all-inclusive set of DataElements. A DataElement alternative, on the other hand, is a DataElement whose value is a list of DataElements of which any one may be selected. Put slightly differently, a DataElement alternative is a set of DataElements of which any one of the values may be used.

The first attribute value in Table 7.1 contains the 32–digit hexadecimal number 68EE141812D211D78EED00B0D03D76EC. This number is also extracted from the connection string. The number is used to create a 128-bit UUID and then is wrapped in a DataElement of type UUID. UUIDs are used extensively in creating service records, and their meaning varies depending on where they are used in the service record. In this case, the UUID represents one of the ServiceClasses in the ServiceClassIDList. ServiceClasses are very important in identifying services. For example, each Bluetooth profile is associated with a particular ServiceClass UUID. If a client wants to find the service record for a particular JABWT server application, it can search for the ServiceClass UUID used by that server application.

The other service class in the ServiceClassIDList is the 16-bit UUID, 0x1101, which identifies this as a serial port service record. This list of two service classes summarizes the type of service being offered. Because this attribute value represents a list of service classes, the two Data-Elements that represent individual service classes are wrapped in another DataElement that represents the entire list (or sequence) of service classes. This wrapper DataElement for the list has type DATSEQ, which is an abbreviation for DataElement Sequence.

In the ServiceClassIDList attribute value in Table 7.1 we see both 16-bit and 128-bit UUIDs. Although a Bluetooth UUID always represents a 128-bit value, the Bluetooth specification defines both 16-bit and 32-bit "short forms" or aliases for some common 128-bit Bluetooth UUIDs. For example, the Serial Port Service Class ID, 0000**1101**0000100080000805F9B34FB, is a 128-bit UUID that has a 16-bit short form 0x1101. It takes fewer bits to store and transmit a service record when these short form UUIDs are used, so the short forms are generally used when they are available. The short-form UUIDs are defined by the Bluetooth Assigned Numbers [4].

A 16-bit UUID can be converted to a 128-bit UUID by means of the following formula:

$$UUID_{128} = (UUID_{16} * 2^{96}) + 0x00000000000001000800000805F9B34FB$$

Table 7.3 shows examples of the use of this formula to convert from a 16-bit UUID to a 128-bit UUID.

The ProtocolDescriptorList attribute value has the most complicated structure of the four attributes shown in Table 7.1. The ProtocolDescriptorList describes how clients can connect to the service described by the service record. It lists the protocol stack needed to communicate with the service and any protocol-specific parameters needed to uniquely address the service. In the example shown in Table 7.1, a

Table 7.3   Examples of Conversion from a 16-Bit UUID to a 128-Bit UUID

Mnemonic	16-Bit UUID	128-Bit UUID
RFCOMM	0x0003	0x0000**0003**00001000800000805F9B34FB
BNEP	0x000F	0x0000**000F**00001000800000805F9B34FB
L2CAP	0x0100	0x0000**0100**00001000800000805F9B34FB
OBEXObjectPush	0x1105	0x0000**1105**00001000800000805F9B34FB

connection to this serial port service can be made by means of a stack of protocols that consists of the L2CAP layer and the RFCOMM layer. The implication is that the server application communicates directly with RFCOMM. Server channel 1 has been assigned to the server application by the Bluetooth stack, and this channel identifier is included in the service record so clients know the proper channel identifier to use to make a connection to the service.

The structure of the ProtocolDescriptorList is a list of lists with one sublist for every stack layer involved in the communications. So conceptually this looks like ((L2CAP), (RFCOMM, 1)), where parentheses are used as shorthand for a DataElement of type DATSEQ. The first element (L2CAP) indicates that L2CAP is the lowest protocol layer used to access this service. Strictly speaking, other Bluetooth stack protocols below L2CAP are involved, but stack layers below L2CAP are not included in SDP service records. The second element, (RFCOMM, 1), consists of two elements. The first element is the name of the next higher layer protocol, RFCOMM. The second element is a protocol-specific parameter, 1, which is the RFCOMM server channel identifier.

The list-of-lists structure is represented in the service record as an attribute value with structure

```
DataElement(type = DATSEQ,
            DataElement(type = DATSEQ, ...)
            DataElement(type = DATSEQ, ...))
```

Short-form UUIDs are used to represent the protocols L2CAP and RFCOMM. A DataElement of type U_INT_1 represents the server channel identifier 1 used by RFCOMM. This type of DataElement describes an unsigned integer of size 1 byte. In addition to the U_INT_1 type, there are DataElement types for signed and unsigned integers that are 1, 2, 4, 8, or 16 bytes long.

The third attribute shown in Table 7.1, ServiceName, has already been discussed. Its value is a DataElement of type String with value "SPPEx." The Bluetooth SDP specification defines the ServiceName attribute as a brief string representing the service suitable for display to the device user.

The fourth service attribute shown in Table 7.1 is the Service-RecordHandle. This is a required attribute for every service record and plays an important role in the implementation of Bluetooth service discovery. However, the ServiceRecordHandle should be considered internal bookkeeping irrelevant to JABWT applications. JABWT applications may not modify this attribute. For this reason, the ServiceRecord-Handle attribute is omitted from the rest of the tables describing service records in this book.

The SDP specification describes the binary format used to transmit (attribute ID, attribute value) pairs from a service record over the air to another Bluetooth device. However, JABWT applications do not see that binary representation. Table 7.1 provides a representation of a service record that maps more directly onto the Java objects visible to JABWT applications. Table 7.1 refers to the JABWT classes `DataElement` and `UUID`.

Service Records for L2CAP

As an alternative to serial port communications using `btspp`, JABWT server applications can communicate using L2CAP or OBEX. Servers that use these protocols need service records different from the one shown for `btspp`. Again, the JABWT implementation is responsible for automatically creating the service records. L2CAP is described in Chapter 8. A server's call to `Connector.open()` using the following `btl2cap` connection string creates a service record like the one shown in Table 7.4:

```
"btl2cap://localhost:BA661F1C148911D783C300B0D03D76EC;
          name=An L2CAP Server"
```

There are several differences between Table 7.1 and Table 7.4:

- SerialPort has been removed from the ServiceClassIDList.
- RFCOMM has been removed from the ProtocolDescriptorList.
- The value of the Protocol/Service Multiplexer (PSM), 0x1001, is included as an L2CAP parameter in the ProtocolDescriptorList.

Because L2CAP servers talk directly to the L2CAP layer in the Bluetooth stack, L2CAP is the last element in the ProtocolDescriptorList that describes the sequence of protocol layers that must be traversed to reach this server application. The protocol and service multiplexer

Table 7.4 A Service Record Created by `Connector.open("btl2cap://localhost:...")`

ServiceClassIDList<0x0001>
DataElement(type = DATSEQ,
DataElement(type = UUID,
UUID(BA661F1C148911D783C300B0D03D76EC)
—from the connection string))
ProtocolDescriptorList<0x0004>
DataElement(type = DATSEQ,
DataElement(type = DATSEQ,
DataElement(type = UUID, UUID(L2CAP<0x0100>))
DataElement(type = U_INT_2, 0x1001
–Protocol/Service Multiplexer. Assigned by
–the stack; filled in by JABWT)))
ServiceName<0x0100>
DataElement(type = STRING, "An L2CAP Server"
–from "name=An L2CAP Server" in the connection string)

parameter is required because L2CAP is a multiplexing layer, so multiple applications may be interacting with the L2CAP layer on the server device. The PSM value in the service record enables L2CAP to identify the particular application or protocol above L2CAP that will provide the service described by the service record. L2CAP uses the PSM value to set up an L2CAP channel to the correct application when a client connects to this service (see Chapter 8).

As was the case for the SPP, the JABWT implementation adds all the mandatory service attributes of the service record for L2CAP. The result is a minimal, but sufficient, service record.

Service Records for OBEX over RFCOMM

A third option for a Bluetooth server application is to use OBEX for communication. The following statement is an example of a connection string used for OBEX over RFCOMM:

```
"btgoep://localhost:0E18AE04148A11D7929B00B0D03D76EC;
        name=An OBEX Server"
```

The abbreviation `goep` in the protocol `btgoep` refers to the Generic Object Exchange Profile, which is the base Bluetooth profile shared by all Bluetooth OBEX profiles. The relation between OBEX and the GOEP profile is similar to the relationship between the RFCOMM protocol and the SPP.

This call to `Connector.open("btgoep://localhost:...")` creates a service record like the one shown in Table 7.5. There are only a few differences between this service record and the one shown in Table 7.1 for `btspp`. In Table 7.5 the service record contains OBEX as the last item in its ProtocolDescriptorList. This indicates that the server application talks directly to the OBEX layer of the Bluetooth stack. Another difference between Table 7.5 and Table 7.1 is that ServiceClassIDList in Table 7.5 does not include the SerialPort service class ID included for `btspp`.

Table 7.6 summarizes the three different protocols used by Bluetooth servers and the `Connector.open()` methods used to create service records for all three protocols. As shown in Table 7.6, the `Connector.open()` method is primarily defined in the CLDC specification. Whereas CLDC provides the primary specification for the behavior of the `Connector.open()` method, the JABWT specification describes the valid `url` arguments for Bluetooth servers and the behavior of `Connector.open()` in creating Bluetooth service records.

Add the Service Record to the SDDB

Although the `Connector.open()` methods in Table 7.6 create a minimal service record for run-before-connect services, that service record is not yet visible to client devices. The server that created the service record can access it and make modifications to it if desired. However, it is

Table 7.5   A Service Record Created by `Connector.open("btgoep://localhost:...")`

ServiceClassIDList<0x0001>

DataElement(type = DATSEQ,

       DataElement(type = UUID,

             UUID(0E18AE04148A11D7929B00B0D03D76EC)

                 —from the connection string))

ProtocolDescriptorList<0x0004>

DataElement(type = DATSEQ,

       DataElement(type = DATSEQ,

             DataElement(type = UUID, UUID(L2CAP<0x0100>)))

             DataElement(type = DATSEQ,

                 DataElement(type = UUID, UUID(RFCOMM<0x0003>))

                 DataElement(type = U_INT_1, 20 –server channel identifier))

             DataElement(type = DATSEQ,

                 DataElement(type = UUID, UUID(OBEX<0x0008>))))

ServiceName<0x0100>

DataElement(type = STRING, "An OBEX Server"

      –from "name=" in the connection string)

---

possible for clients of run-before-connect services to connect only after the server calls `acceptAndOpen()`. For this reason, the JABWT implementation adds a service record to the SDDB only the first time the server calls one of the `acceptAndOpen()` methods in Table 7.7. Once the service record is in the SDDB, client applications can discover that service record and attempt to connect to the server application.

A RFCOMM service can accept multiple connections from different clients by calling `acceptAndOpen()` repeatedly for the same notifier.

Table 7.6   Methods That Create a Service Record

Protocol	Interface	Methods	Specifications
btspp	Connector	open(url) open(url, mode) open(url, mode,timeout)	CLDC
btl2cap	Connector	open(url) open(url, mode) open(url, mode,timeout)	CLDC
btgoep	Connector	open(url) open(url, mode) open(url, mode,timeout)	CLDC

Table 7.7   Methods That Add Service Records to the SDDB

Protocol	Interface	Methods	Specification
btspp	StreamConnectionNotifier	acceptAndOpen()	CLDC
btl2cap	L2CAPConnectionNotifier	acceptAndOpen()	JABWT
btgoep	SessionNotifier	acceptAndOpen(handler) acceptAndOpen(handler, authenticator)	JABWT

Each client accesses the same service record and connects to the service using the same RFCOMM server channel. If the underlying Bluetooth system does not support multiple connections, then the implementation of acceptAndOpen() throws a BluetoothStateException. L2CAP and OBEX over RFCOMM services also can accept multiple clients.

A ServiceRegistrationException is thrown by all of the acceptAndOpen() methods in Table 7.7 if they fail to add a service record to the SDDB.

Table 7.8   Methods That Remove or Disable Service Records

Protocol	Interface	Methods	Specification
btspp	StreamConnectionNotifier	close()	CLDC
btl2cap	L2CAPConnectionNotifier	close()	CLDC
btgoep	SessionNotifier	close()	CLDC

Remove the Service Record from the SDDB

Once the notifier associated with a run-before-connect service is closed, it is no longer possible to call acceptAndOpen() to accept another client connection. For this reason, the JABWT implementation removes the service record from the SDDB or disables the service record. Table 7.8 shows the different types of notifiers that add this behavior to the close() method inherited from the GCF interface javax.microedition.io.Connection.

7.2.3   Modifications to Service Records

In many cases, it is desirable to modify the service record created by the JABWT implementation. For example, if your service corresponds to a Bluetooth profile, you will have to modify the service record so that the record conforms to the requirements of the profile. Even if you are writing a custom application and are not required to have a standardized service record, you may want to modify your service record to provide various kinds of useful information to potential clients. Many optional attributes are defined in the Bluetooth SDP specification that server applications can use to describe the properties of their service. It is also possible to add application-specific, user-defined attributes to the service record that are not defined by the Bluetooth specification.

Figure 7.3 adds JABWT methods for modifying service records to the sequence diagram shown in Figure 7.2. The LocalDevice class provides a getRecord() method that a server application can use to obtain its ServiceRecord. The server can modify the ServiceRecord

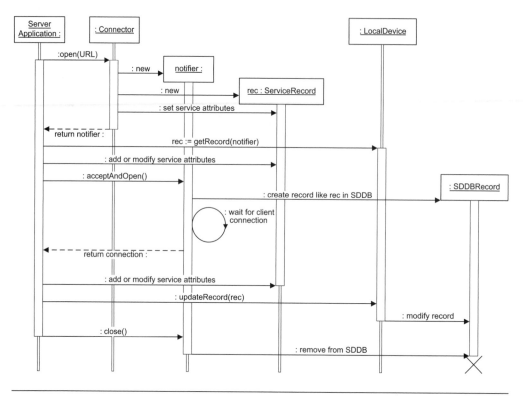

Figure 7.3   Example of a server modifying its service record.

object by adding or modifying attributes using `ServiceRecord.`
`setAttributeValue()`. As shown in Figure 7.3, any modifications
the server application makes to its `ServiceRecord` before calling
`acceptAndOpen()` for the first time will be reflected in the service
record added to the SDDB by `acceptAndOpen()`.

Any changes made to the service record object by a JABWT application after the first call to `acceptAndOpen()` are not reflected in the
service record in the SDDB seen by clients. This is because the service
record in the SDDB is essentially a copy of the service record Java object
at the time of the first call to `acceptAndOpen()`. To modify service
records already in the SDDB, JABWT provides the instance method
`LocalDevice.updateRecord(serviceRecord)`.

7.2.4 Device Service Classes

As described in Chapter 6, clients can consult the `DeviceClass` of any device they discover to determine what kind of device has been found (e.g., phone, PDA, or PC). The `DeviceClass` also indicates the major service classes offered by the discovered device (e.g., rendering, telephony, or information). This means there are two different ways in which a server application describes the service it offers:

- By adding a service record to the SDDB
- By activating major service class bits in the `DeviceClass`

The server application can use the `setDeviceServiceClasses()` method of the `ServiceRecord` class to turn on some of the service class bits of the device to reflect the new service being offered. A server application is not required to use the `setDeviceServiceClasses()` method. However, it is recommended that a server use the method to describe its service in terms of the major service classes. Keeping the major service classes up to date reduces the likelihood that clients will erroneously skip over this device when looking for a service.

The `close()` message also causes the JABWT implementation to deactivate any service class bits that were activated by `setDevice-ServiceClasses()`, unless another service whose notifier is not yet closed also activated some of the same bits.

7.3 Programming with the API

The programming examples in this chapter are divided into examples of service registration and examples of service discovery. Sections 7.3.1 through 7.3.4 provide examples of service registration. These sections show examples of the use of methods for creating and modifying service records. The examples in these sections are all server applications. These servers simply create a service record and add it to the SDDB. No client applications are needed to illustrate this behavior. Not all of the code needed to produce a running application is presented in the text. The complete code is available on the book's Web site located at www.mkp.com. Sections 7.3.5 through 7.3.8 provide examples of service discovery. These sections extend the `DiscoveryMIDlet` that was introduced in Chapter 6 to discover various aspects of the service defined in Section 7.3.2.

7.3.1 Automatic Generation of Service Records

In this first example, the server application makes no modifications to the service record. This is the simplest case. Figure 7.4 shows the output produced by the `DefaultBtsppRecordMIDlet`. The display shows the connection string that clients can use to connect to this server:

btspp://002233445566:1

The display also shows that the service record for the server has five service attributes and lists their attribute IDs as hex numbers.

The `DefaultBtsppRecordMIDlet` implements the `Runnable` interface. The `run()` method first calls the method `askToBeGeneral-Discoverable()` defined in the `DefaultBtsppRecordServer` class to attempt to make the server device general discoverable. The `run()` method calls the method `defineDefaultBtsppService()` to create the service record and create the `StreamConnectionNotifier`. The new service record is obtained from the `LocalDevice`, and a brief

Figure 7.4   Example code displays information about the default service record (emulation only).

description of the service record is appended to a `Form`. Finally, the `run()` method calls the `acceptClientConnections()` method defined in the `DefaultBtsppRecordServer` class. This method adds the service record to the SDDB and waits for client connections.

```
public class DefaultBtsppRecordMIDlet extends MIDlet
  implements Runnable, CommandListener {

  StreamConnectionNotifier notifier;

  /* The form displayed to the user. */
  private Form output;

  ...

  public void run() {
    LocalDevice theRadio;
    // Define the serial port service and create the notifier
    try {
      theRadio = LocalDevice.getLocalDevice();
      server = new DefaultBtsppRecordServer();
      server.askToBeGeneralDiscoverable(theRadio);
      notifier = server.defineDefaultBtsppService();
    } catch (IOException e) {
      output.append("Unable to start server (IOException: " +
        e.getMessage() + ")");
      return;
    }

    if (notifier != null) {
      ServiceRecord record = theRadio.getRecord(notifier);
      output.append("URL=" + server.getURL(record));
      output.append(server.describeAttributes(record));
    } else {
      output.append("Unable to start server");
      return;
    }
    // Use the notifier to establish serial port connections
    server.acceptClientConnections(notifier);
  }
}
```

Now that we have seen the overall flow of execution defined by the
`DefaultBtsppRecordMIDlet`, we will examine the `DefaultBtspp-`
`RecordServer` class. The `askToBeGeneralDiscoverable()` method
uses the `setDiscoverable()` method to request that the device be made
general discoverable. This enables client devices that do device discovery
with the GIAC mode to find the server device. If `setDiscoverable()`
returns `false`, indicating that the request was not granted, or if it throws
an exception, the server just proceeds. Any clients that know the Bluetooth address for this server can access this service even if the device is not
discoverable. For example, clients that include the server device among
their pre-known devices can access the server (see Chapter 6).

The `defineDefaultBtsppService()` method calls `Connector.`
`open(connString)` to create a `StreamConnectionNotifier`. That
same call also creates a default `btspp` service record such as the one
shown in Table 7.1 and associates it with the notifier.

```java
public class DefaultBtsppRecordServer {
  boolean stop = false;
  void askToBeGeneralDiscoverable(LocalDevice dev) {
    try {
      /* Request that the device be made discoverable */
      dev.setDiscoverable(DiscoveryAgent.GIAC);
    } catch(BluetoothStateException ignore) {
      /* discoverable is not an absolute requirement */
    }
  }

  public StreamConnectionNotifier
    defineDefaultBtsppService() {
    StreamConnectionNotifier notifier;
    String connString = "btspp://localhost:" +
      "68EE141812D211D78EED00B0D03D76EC;" +
      "name=SPPEx";
    try {
      notifier = (StreamConnectionNotifier)
        Connector.open(connString);
    } catch (IOException e){
      return null;
```

```
    }
    return notifier;
}
public String getURL(ServiceRecord record) {
    String url = record.getConnectionURL(
        ServiceRecord.NOAUTHENTICATE_NOENCRYPT,
        false);
    if (url != null) {
        return url.substring(0, url.indexOf(";"));
    } else {
        return "getConnectionURL()=null";
    }
}
public String describeAttributes(ServiceRecord
    record) {
    int[] attributeIDs = record.getAttributeIDs();
    StringBuffer strBuf = new StringBuffer(100);
    strBuf.append("\n").append(Integer.toString(
        attributeIDs.length));
    strBuf.append(" Attributes: ");
    for (int i = 0; i < attributeIDs.length; i++) {
        strBuf.append("<0x");
        strBuf.append(Integer.toHexString(attributeIDs[i]));
        strBuf.append(">\n");
    }
    return strBuf.toString();
}
public void acceptClientConnections(
    StreamConnectionNotifier notifier) {
    if (notifier == null) {
        return;
    }
    try {
        while (!stop) {
```

```
      StreamConnection clientConn = null;
      /*
       * acceptAndOpen() waits for the next client to
       * connect to this service. The first time through
       * the loop, acceptAndOpen() adds the service record
       * to the SDDB and updates the service class bits
       * of the device.
       */
      try {
        clientConn =
          (StreamConnection)notifier.acceptAndOpen();
      } catch (ServiceRegistrationException e1) {
      } catch (IOException e) {
        continue;
      }

      /*
       * Code to communicate to a client over clientConn
       * would go here.
       */
    }
  } finally {
    try {
      shutdown(notifier);
    } catch (IOException ignore) {
    }
  }
}

public void shutdown(StreamConnectionNotifier notifier)
  throws IOException {
  stop = true;
  notifier.close();
}
}
```

The getURL() method returns a connection string that clients can use to connect to the DefaultBtsppRecordServer. The getURL() method calls the JABWT getConnectionURL() method to get the connection string, and then the string is shortened for display by removing the parameter list. As shown in Figure 7.4, the result is `btspp://002233445566:1`, where `002233445566` is the Bluetooth address of the local device, and 1 is the server channel identifier. Typically, clients send the `getConnectionURL()` message to a service record obtained during service discovery to obtain a connection string to connect to that service. Here we send the same message to the server's own service record to obtain the connection string for display by the `DefaultBtsppRecordMIDlet`.

The `describeAttributes()` method uses the JABWT method `getAttributeIDs()` to obtain an array of the attribute IDs that are part of the new service record. The `describeAttributes()` method returns a string that includes the number of attributes in this array and the hexadecimal values of the attribute IDs. The `DefaultBtsppRecordMIDlet` displays this string on the user interface. These attribute IDs can be compared with the ones shown for the default `btspp` service record in Table 7.1. The `DefaultBtsppRecordMIDlet` displays the attribute IDs in Table 7.1. (Some JSR-82 implementations do not return a ServiceRecordHandle, 0x0000.) One additional attribute, ServiceRecordState 0x002, might also be displayed. The Bluetooth stack may add the ServiceRecordState attribute to a service record to make it easier for clients to determine whether that service record has changed. If the value of the ServiceRecordState attribute has not changed since the last time it was checked, the client knows that none of the attributes in the service record have changed.

The last method defined in `DefaultBtsppRecordServer` is `acceptClientConnections()`. This method calls `acceptAndOpen()`, which adds the service record to the SDDB, where it will be visible to clients. The `acceptAndOpen()` method then blocks and waits for a client to connect. Once a client makes a connection, the `acceptAndOpen()` method returns a `StreamConnection` that the server can use to communicate with that client using RFCOMM (see Chapter 4).

7.3.2 Modifying a Service Record

This section illustrates how a server can modify its service record by adding additional service attributes. Suppose we want to create the service record shown in Table 7.9 for a two-person Bluetooth game.

Table 7.9   The Service Record for a Bluetooth Game

ServiceClassIDList<0x0001>

DataElement(type = DATSEQ,
        DataElement(type = UUID,
                UUID(0FA1A7AC16A211D7854400B0D03D76EC))
        DataElement(type = UUID, UUID(SerialPort<0x1101>)))

ProtocolDescriptorList<0x0004>

DataElement(type = DATSEQ,
        DataElement(type = DATSEQ,
                DataElement(type = UUID, UUID(L2CAP<0x0100>)))
        DataElement(type = DATSEQ,
                DataElement(type = UUID, UUID(RFCOMM<0x0003>))
                DataElement(type = U_INT_1, 3 –server channel identifier)))

ServiceName<0x0100>
DataElement(type = STRING, "A Bluetooth Game")
ServiceDescription<0x0101>
DataElement(type = STRING,   "This game is fun! It is for two people. You can play it on your cell phones.")
DocumentationURL<0x000A>–Where to find documentation
DataElement(type = URL, "http://www.gameDocsOnSomeWebPage.com")
<0x2222>–An application-specific attribute for the highest score in the game
DataElement(type = U_INT_4, 10000)

The JABWT implementation automatically adds the first three attributes shown in Table 7.9 when it creates the service record. The last three attributes must be added by the server application. Two of the service attributes added, ServiceDescription and DocumentationURL, are standard attributes defined in the SDP specification. A ServiceDescription is a brief description of the service (fewer than 200 characters). The DocumentationURL provides a pointer to a Web page for detailed documentation of the service. The third attribute added, 0x2222, is a nonstandard, application-specific service attribute. This attribute shows the highest score achieved to date by the user of this device. Clients might use this attribute to select a suitable opponent for the game or to assign handicaps.

The defineGameService() method shown below illustrates how the service record shown in Table 7.9 can be created by a server application. The statement

```
notifier = (StreamConnectionNotifier)
  Connector.open(connString)
```

creates the service record and returns a notifier cast to a Stream-ConnectionNotifier. The notifier is used to access the new service record by the statement

```
ServiceRecord record = localDev.getRecord(notifier);
```

The defineGameService() method then adds three additional service attributes to the service record before that record is made visible to clients. The method setAttributeValue() is used to add each attribute to the service record.

```
public StreamConnectionNotifier
  defineGameService( LocalDevice localDev, long highScore) {
  StreamConnectionNotifier notifier;
  String connString =
    "btspp://localhost:0FA1A7AC16A211D7854400B0D03D76EC;"+
    "name=A Bluetooth Game";
  try {
  notifier =
```

```
      (StreamConnectionNotifier)Connector.open(connString);
  } catch (IOException e2){
    return null;
  }
  ServiceRecord record = localDev.getRecord(notifier);
  // Add optional ServiceDescription attribute;
  // attribute ID 0x0101.
  record.setAttributeValue(0x0101,
    new DataElement(DataElement.STRING,
    "This game is fun! It is for two people. " +
    "You can play it on your cell phones."));
  // Add optional DocumentationURL attribute; attribute ID
  // 0x000A.
  record.setAttributeValue(0x000A,
    new DataElement(DataElement.URL,
    "http://www.gameDocsOnSomeWebpage.com"));
  /*
   * Add an application-specific attribute for the highest
   * score achieved by this player to date.
   */
  record.setAttributeValue(0x2222,
    new DataElement(DataElement.U_INT_4,
    highScore));
  return notifier;
}
```

When the server does `acceptAndOpen()`, the service record added to the SDDB has the additional service attributes added by the `define-GameService()` method. When using application-specific service attributes, keep in mind that the Bluetooth specification reserves certain attribute ID values. Attribute IDs in the range 0x000D to 0x01FF are reserved and should not be used.

7.3.3   Support for String Attributes in Several Languages

Table 7.10 shows how Bluetooth service records can include strings in more than one language. In addition to the ServiceName and Service-Description attributes shown in Table 7.9, three service attributes have

Table 7.10 Selected Attributes From a Service Record with English and French Strings

...
ServiceName<0x0100>
–Name of the service in the primary language of the service record
DataElement(type = STRING, "A Bluetooth Game"
–from "name=" in the connection string)
ServiceDescription<0x0101>
–Description of the service in the primary language
DataElement(type = STRING,
"This game is fun! It is for two people. You can play it on your cell phones.")
ServiceName<0x0120> –Name of the service in French
DataElement(type = STRING, "Jeu de Bluetooth")
ServiceDescription<0x0121> –Description of the service in French
DataElement(type = STRING,
"Ce jeu est amusant ! Il se joue à deux. Vous pouvez y jouer sur vos téléphones mobiles.")
LanguageBaseAttributeIDList<0x0006>
–Describe the languages used in the service record
DataElement(type = DATSEQ,
DataElement(type = U_INT_2, 0x656E –ASCII for "en", English)
DataElement(type = U_INT_2, 0x006A –the MIBenum for UTF-8)
DataElement(type = U_INT_2, 0x0100 –attribute ID base for English)
DataElement(type = U_INT_2, 0x6672 –ASCII for "fr", French)
DataElement(type = U_INT_2, 0x006A –the MIBenum for UTF-8)
DataElement(type = U_INT_2, 0x0120 –attribute ID base for French))

been included in the service record. Two of these attributes provide the ServiceName and ServiceDescription in French. The third attribute, the LanguageBaseAttributeIDList, describes the two languages used in this service record and provides the information needed to differentiate the English strings from the French strings.

To support use of multiple languages in service records, the Bluetooth SDP uses a base-plus-offset scheme for all service attributes of type string. In the service record shown in Table 7.10, the base for English service attributes is 0x0100. The base for French service attributes is 0x0120. The SDP specification defines the ServiceDescription as having an offset of 0x0001. This means the attribute ID of the ServiceDescription in English in this service record is given by the English base plus offset; or

0x0100 + 0x0001 = 0x0101

The attribute ID of the ServiceDescription in French is given by the French base plus offset, or

0x0120 + 0x0001 = 0x0121

ServiceName is defined as having an offset of 0x0000, so the ServiceName in English has attribute ID

0x0100 + 0x0000 = 0x0100

The ServiceName in French has attribute ID

0x0120 + 0x0000 = 0x0120

The LanguageBaseAttributeIDList contains the attribute ID base for each language. LanguageBaseAttributeIDList is an optional service attribute. However, if a service record were to use more than one language, it would be very difficult for applications to use the other languages without knowing the attribute ID base value for the other languages. The attribute value for the LanguageBaseAttributeIDList is a list, or DATSEQ, of DataElements of type U_INT_2, that is, of type unsigned 2-byte integer.

The elements of the LanguageBaseAttributeIDList are implicitly grouped into triplets, where each triplet is for a particular language. The first element of the triplet is the language code as standardized by ISO 639

[36]. For English, this code is "en"; for French, it is "fr." The second element of the triplet is the character encoding used for the language. Unicode Transformation Format 8 (UTF-8) is an example of character encoding. The Internet Assigned Numbers Authority maintains a standard list of character encodings [37]. Each encoding in the standard has a MIBenum value; for example, UTF-8 is 0x006A (decimal 106). The MIBenum value of the character encoding is the second element in each triplet contained in the LanguageBaseAttributeIDList. The third element of each triplet is the attribute ID base value for the triplet's language.

The base values used in the third element of each triplet are not standardized. The service records in an SDDB may use different base values for the same language. The only rules are the following:

- The attribute ID base for the primary language used in a particular service record must be 0x0100.
- If there is a LanguageBaseAttributeIDList in the service record, the first language in this list must use the attribute ID base 0x0100; that is, the first language must be the primary language.

The attribute ID base used for a second language is not standardized; that is, we chose the value of 0x0120 for the base value for French in Table 7.10. In selection of base values, care must be taken to avoid conflicts with the attribute IDs defined by the Bluetooth Assigned Numbers. One recommendation is to choose attribute ID base values so that the sum of base value plus offset falls either in the range 0x0100 to 0x01FF or in the range 0x0900 to 0xFFFF. The same recommendation applies to user-defined, service-specific attributes with string values.

7.3.4 Service Records for Bluetooth Profiles

The Bluetooth profiles describe a number of common tasks that will be accomplished with Bluetooth wireless technology. The profiles list requirements that help in achieving interoperability between devices with independent implementations of these standardized tasks. Developers who intend to implement one of these Bluetooth profiles with their JABWT application need to study the specifications closely so that their applications can pass any Bluetooth qualification tests for the profile and can successfully interoperate with other devices that also support the profile.

The profile specifications place requirements on both client and server roles for each profile. Part of the requirements for servers is a specification of what the service record will look like for the profile service. As an example, Table 7.11 shows the service record for the Object Push Profile. This format is followed by all of the Bluetooth profiles.

The Object Push Profile describes how electronic business cards and other similar objects can be transmitted between Bluetooth devices by means of OBEX. The server in the Object Push Profile is called a Push Server, and the client is called a Push Client. As Table 7.11 shows, the UUID with short form 0x1105 is defined as the Service Class ID for Push Servers that conform to the Object Push Profile. Push Clients can look for service records that contain this UUID to identify Push Servers that have demonstrated they meet the requirements of the Object Push Profile.

The Status column in Table 7.11 indicates service record entries that are mandatory (Mand.) and entries that are optional (Opt.) according to the Bluetooth specification.

The UUID values in the Value and Default Value columns shown in Table 7.11 are obtained from the Bluetooth Assigned Numbers [34]. The service attribute ID values in the AttrID column also are obtained from the Bluetooth Assigned Numbers. The version of Table 7.11 in the Object Push Profile specification does not contain these values explicitly but instead refers to the assigned numbers.

The service record in Table 7.11 can be translated into the same notation used earlier in this chapter to describe the `btspp`, `btl2cap`, and `btgoep` service records. That translation is shown in Table 7.12. The translation is straightforward for the most part. However, the two notations use different units (bits versus bytes) when describing types for numbers. Table 7.11 from the Bluetooth profiles specification uses `Uint8` and `Uint16`, which refer to unsigned 8-bit and 16-bit integers. Table 7.12, on the other hand, uses `U_INT_1` and `U_INT_2` for these same two quantities. Table 7.12 uses the names of JABWT constants for all the "type =" entries. The `javax.bluetooth.DataElement` constants `U_INT_1` and `U_INT_2` refer to 1-byte and 2-byte integers. The representations in Table 7.11 and Table 7.12 are equivalent but use different units (bits versus bytes) when describing the type of a number.

Table 7.11   Service Record Defined by the Bluetooth Object Push Profile

Item	Definition	Type	Value	AttrID	Status	Default Value
ServiceClassIDList				0x0001	Mand.	
ServiceClass #0	UUID for OBEXObjectPush	UUID	0x1105		Mand.	0x1105
ProtocolDescriptorList				0x0004	Mand.	
Protocol ID #0	UUID for L2CAP protocol	UUID	0x0100		Mand.	0x0100
Protocol ID #1	UUID for RFCOMM protocol	UUID	0x0003		Mand.	0x0003
Parameter #0	Server channel	UINT8	varies		Mand.	varies
Protocol ID #2	UUID for OBEX	UUID	0x0008		Mand.	0x0008
ServiceName	Displayable Text name	String	varies	0x0000 + base	Opt.	"OBEX Object Push"
BluetoothProfile DescriptorList				0x0009	Opt.	
Profile ID #0	Supported profile	UUID	0x1105			0x1105
Version #0	Profile version	UINT16	0x0100			0x0100
Supported Formats List	Supported Formats List	Data Element Sequence of UINT8	Formats: 0x01 = vCard 2.1 0x02 = vCard 3.0 0x03 = vCal 1.0 0x04 = iCal 2.0 0x05 = vNote 0x06 = vMessage 0xFF = any type of object	0x0303	Mand.	

Table 7.12   Service Record for an OBEX Object Push Server

ServiceClassIDList<0x0001>

DataElement(type = DATSEQ,
        DataElement(type = UUID, UUID(OBEXObjectPush<0x1105>)
                —Object Push UUID))

ProtocolDescriptorList<0x0004>

DataElement(type = DATSEQ,
        DataElement(type = DATSEQ,
                DataElement(type = UUID, UUID(L2CAP<0x0100>)))
        DataElement(type = DATSEQ,
                DataElement(type = UUID, UUID(RFCOMM<0x0003>))
                DataElement(type = U_INT_1, 4 –server channel identifier))
        DataElement(type = DATSEQ,
                DataElement(type = UUID, UUID(OBEX<0x0008>))))

ServiceName<0x0100>
DataElement(type = STRING, "OBEX Object Push Server")
BluetoothProfileDescriptorList<0x0009>

DataElement(type = DATSEQ,
        DataElement(type = DATSEQ,
                DataElement(type = UUID, UUID(OBEXObjectPush<0x1105>))
                DataElement(type = U_INT_2, 0x0100 – version 1.00)))

Supported Formats List<0x0303>

DataElement(type = DATSEQ,
        DataElement(type = U_INT_1, 0x01 – vCard 2.1)
        DataElement(type = U_INT_1, 0x02 – vCard 3.0))

There are several points to observe about Table 7.12. In addition to the familiar attributes ServiceClassIDList, ProtocolDescriptorList, and ServiceName are several attributes that we have not seen before. Table 7.12 includes a BluetoothProfileDescriptorList, which is an attribute used to declare that this service conforms to version number 1.00 of the Object Push Profile. The service records for Bluetooth profiles commonly include a BluetoothProfileDescriptorList attribute, although that attribute usually is optional. The short-form UUID for OBEXObjectPush, 0x1105, is used in the BluetoothProfileDescriptorList to designate the Object Push Profile. Because this UUID also is used in the ServiceClassIDList, this marks the second appearance of this same UUID in the service record.

Table 7.12 also includes one attribute, Supported Formats List, that is unique to the Object Push Profile. The Supported Formats List attribute describes the different object formats supported by this Push Server. The Supported Formats List entry in Table 7.11 describes the different object formats recognized by the Object Push Profile. In Table 7.12, the two formats for electronic business cards have been specified. The Object Push Profile requires that Phone Book applications support data exchange using the vCard 2.1 content format [38].

The following example code shows how a server application can create an Object Push service record. After `Connector.open()` is used to create the service record, the `defineObjectPushService()` method adds the BluetoothProfileDescriptorList attribute to the service record by calling the application method `setBluetoothProfileList()`. The code for this method is presented later. It is followed by code that adds the Supported Formats List attribute to the service record to declare that this server understands the vCard 2.1 and 3.0 formats.

```
/**
 * Create the service record for an OBEX Object Push
 * server as defined by the Bluetooth Object Push
 * profile.
 */
public SessionNotifier defineObjectPushService() {
  SessionNotifier notifier;
  // The UUID 00001105000... is the long-form UUID for the
  // short form 0x1105 defined for the Object Push
```

```
// service ID by assigned numbers.
String connString = "btgoep://localhost:" +
  "000011050000100080000805F9B34FB;" +
  "name=OBEX Object Push Server";

// Connector.open() assigns a RFCOMM server channel
// and creates a service record using this channel.
try {
  notifier =
    (SessionNotifier)Connector.open(connString);
} catch (ServiceRegistrationException e1) {
  // The open method failed because unable to obtain
  // an RFCOMM server channel.
  return null;
} catch (IOException e2){
  // The open method failed due to another IOException
  return null;
}

try {
  localDev = LocalDevice.getLocalDevice();
} catch (BluetoothStateException e) {
  return null;
}

ServiceRecord record = localDev.getRecord(notifier);

// Add the optional service attribute
// BluetoothProfileDescriptorList
ServiceRecordUtilities.setBluetoothProfileList(record,
  0x1105, 0x0100);
DataElement objFormatsDE = new
  DataElement(DataElement.DATSEQ);

// supported format 0x01 = vCard 2.1
objFormatsDE.addElement(new
  DataElement(DataElement.U_INT_1, 0x01));

// supported format 0x02 = vCard 3.0
objFormatsDE.addElement(new
  DataElement(DataElement.U_INT_1, 0x02));
```

```
// Add mandatory Supported Formats List, attribute ID
// 0x0303
record.setAttributeValue(0x0303, objFormatsDE);
// An Object Push Server provides an Object Transfer
// service.
// Bit 20 of the Class of Device is for Object Transfer.
record.setDeviceServiceClasses(0x100000);

return notifier;
}
```

Table 7.12 shows that the Object Push service record has the following ServiceClassIDList:

```
DataElement(type = DATSEQ,
              DataElement(type = UUID,
                          UUID(OBEXObjectPush<0x1105>)
                              —Object Push UUID))
```

This step is accomplished by the method defineObject-PushService() by using the connection string:

```
String connString = "btgoep://localhost:" +
    "0000110500001000800000805F9B34FB;" +
    "name=OBEX Object Push Server";
```

The JABWT implementation inserts the UUID from the connection string into the ServiceClassIDList of the service record. Although there is a short form UUID of 0x1105 for the OBEXObjectPush service class, we have used the long form in the connection string. The reason is that a UUID in a connection string is always interpreted as a 128-bit UUID. This means that 1105 would be interpreted as the 128-bit UUID 00000000000000000000000000001105 not as the 128-bit UUID 0000110500001000800000805F9B34FB.

In the example code, the defineObjectPushService() method uses the setDeviceServiceClasses() method of the Service-Record interface to describe the object transfer major service class provided by the server application

```
record.setDeviceServiceClasses(0x100000);
```

The specification of the Object Push Profile requires that the Push Server indicate that it offers this object transfer service in its device class. A server uses the `setDeviceServiceClasses()` method to associate the `ServiceRecord` with all of the major service classes that describe that service. Later, when a server first calls `acceptAndOpen()`, both its service record and its major service class bits are made visible to client devices. In the case of the major service classes, `acceptAndOpen()` performs a logical OR of the current settings of the service class bits of the device with the major service classes declared by the `setDevice-ServiceClasses()` method. This OR operation might activate additional service class bits that indicate new capabilities for the device.

The `defineObjectPushService()` method makes a static method call to create the BluetoothProfileDescriptorList. The code for that static method is shown below. It creates the DataElement structure required for the BluetoothProfileDescriptorList attribute by the Object Push Profile. This method uses the `DataElement.addElement()` method to assemble DataElements of type `DATSEQ`. One `DataElement` of type `DATSEQ` is used for the pair of profile and version number. Another `DataElement` of type `DATSEQ` is used for the list of pairs. In this case, there is only one pair in the list of pairs (Table 7.12).

```
public class ServiceRecordUtilities {
  public static final int ATT_ID_BLUETOOTH_PROFILE_LIST =
    0x0009;
  /**
   * Sets the value of the BluetoothProfileDescriptorList
   * attribute to be the profile represented by a
   * short-form UUID value and version number.
   * @param record The service record to be modified
   * @param profileUuidValue The short-form UUID for the
   * profile from the Bluetooth Assigned Numbers
   * @param version The version of the profile this
   * service conforms to. The format is 0xMMmm where MM
   * is the major version number and mm is the minor
   * version number.
   */
```

```
public static void setBluetoothProfileList(
  ServiceRecord record,
  long profileUuidValue,
  int version) {
  UUID profileUuid = new UUID(profileUuidValue);
  DataElement profileUuidDE = new
    DataElement(DataElement.UUID,
    profileUuid);
  DataElement versionDE = new
    DataElement(DataElement.U_INT_2, version);
  DataElement profileVersionPairDE = new
    DataElement(DataElement.DATSEQ);
  DataElement profileDescriptorDE = new
    DataElement(DataElement.DATSEQ);
  // Create a pair with profile UUID and profile version
  profileVersionPairDE.addElement(profileUuidDE);
  profileVersionPairDE.addElement(versionDE);
  // Add the pair to the list of profiles
  profileDescriptorDE.addElement(profileVersionPairDE);
  // Set the BluetoothProfileDescriptorList to a DATSEQ data
  // element containing the UUID-version pair for this
  // profile.
  record.setAttributeValue(ATT_ID_BLUETOOTH_PROFILE_LIST,
    profileDescriptorDE);
  }
}
```

7.3.5   Service Discovery

Service discovery within Bluetooth wireless technology can be as complicated or as simple as desired. Like an inquiry, service discovery is a
non-blocking request. As service records are discovered, they are passed
to the application as events. Also, like an inquiry, an event occurs at the
end of the service search to notify the application that the service search
has been completed. Unlike an inquiry, many devices support multiple
service searches at any one time. The number of service searches that the

local device supports can be retrieved via a Bluetooth device property. The "bluetooth.sd.trans.max" property can be retrieved via the `Local-Device.getProperty()` method to determine the maximum number of concurrent service searches.

For the local device to search for a service on a remote device, the local device sends a list of UUIDs to search for to the remote device. The remote device checks all the service records on it for all of the UUIDs sent to it. For every service record that has all the UUIDs, the remote device sends back the ServiceRecordHandle and the requested attributes for that service record.

Starting a service search on a remote device begins with the `DiscoveryAgent.searchServices()` method. The `searchServices()` method takes four arguments. The first argument is the list of attributes to retrieve in any service record that meets the other search criteria. By default, the `searchServices()` method retrieves the ServiceRecord-Handle, ServiceClassIDList, ServiceRecordState, ServiceID, and Protocol-DescriptorList attributes. These attributes are known as the *default attributes*. If the attributes list argument is `null`, only the default attributes are retrieved. If a list of attributes is provided, the default attributes are retrieved in addition to the list provided. With the default attributes, the application has enough information to establish a connection to the service. Additional attributes may be retrieved if additional information about the service is needed. The second argument, the list of UUIDs to search for, specifies all the UUIDs that must exist in a service to be retrieved. The more complete this list of UUIDs, the less likely it is that a service record will contain all of these UUIDs. The third argument, the remote device to search, is a `RemoteDevice` object received via an inquiry or a call to `retrieveDevices()`. The final argument is the `DiscoveryListener` object that will be notified when the services are discovered.

The `searchServices()` method returns the transaction ID for the service search if the device is able to start the service search. The transaction ID allows an application to cancel the search, identify which search located a service, and determine when a specific search is completed. The `searchServices()` method may throw a `BluetoothStateException` if the local device has reached the maximum number of service searches or if the current service search could not be started.

As services are discovered, they are sent to the `DiscoveryListener` via the `servicesDiscovered()` method. The transaction ID of the service search along with all the service records found during the search are provided as arguments. The `servicesDiscovered()` method can be called multiple times for a single service search request. The service records are returned as an array of `ServiceRecord` objects. Each of these `ServiceRecord` objects contains all the attributes requested in the call to `searchServices()` along with the default attributes.

When the service search is completed, the `serviceSearch-Completed()` method is called. The `serviceSearchCompleted()` method provides the transaction ID of the search that ended and a completion status code. Table 7.13 lists all the completion status codes and what those codes mean.

A service search can be canceled with the `cancelService-Search()` method. The `cancelServiceSearch()` method takes as an argument the transaction ID of the service search to cancel. The method returns `true` if the search was canceled. Canceling the search also causes a `serviceSearchCompleted()` event to occur with the `SERVICE_SEARCH_TERMINATED` status code. If the method returns `false`, either the service search has already ended or the transaction ID is not valid.

Table 7.13 Status Codes for Service Searches

Completion Status	Reason
`SERVICE_SEARCH_COMPLETED`	At least one service record was found and the search completed normally.
`SERVICE_SEARCH_TERMINATED`	The service search was canceled by a call to `cancelServiceSearch()`.
`SERVICE_SEARCH_ERROR`	An error occurred during the service search.
`SERVICE_SEARCH_NO_RECORDS`	No records were found during the service search.
`SERVICE_SEARCH_DEVICE_NOT_REACHABLE`	The `RemoteDevice` specified to `searchServices()` could not be reached (i.e., a connection could not be established to the remote device).

Returning to the `DiscoveryMIDlet` introduced in Chapter 6, the next step is to search for services. The `DiscoveryMIDlet` will be modified to search for the Bluetooth game service defined earlier in this chapter. To determine which device to search, the `DiscoveryMIDlet` waits until the user selects a Bluetooth device from a `List`. The `DiscoveryMIDlet` then searches the device specified for the UUID defined by the Bluetooth game service. After retrieving all the services that use this UUID, the `DiscoveryMIDlet` displays the name of each service. Before starting the service search, the `DiscoveryMIDlet` must be modified to maintain a list of the `RemoteDevice` objects found via device discovery and to keep track of when the device is in an inquiry.

```java
public class DiscoveryMIDlet extends BluetoothMIDlet
  implements DiscoveryListener {

    ...
```

```java
    /**
     * Keeps track of the RemoteDevice objects.
     */
    private Vector deviceVector;
    /**
     * Specifies if an inquiry is currently occurring.
     */
    private boolean isInInquiry;
```

```java
    ...
    public void startApp() throws MIDletStateChangeException {
```

```java
      isInInquiry = false;
```

```java
      // Create a new List and set it to the current
      // displayable
      deviceList = new List("List of Devices", List.IMPLICIT);
      deviceList.addCommand(new Command("Exit",
        Command.EXIT, 1));
      deviceList.setCommandListener(this);
```

```
Display.getDisplay(this).setCurrent(deviceList);
// Retrieve the DiscoveryAgent object. If the
// retrieving the local device causes a
// BluetoothStateException, something is wrong
// so stop the app from running.
try {
  LocalDevice local = LocalDevice.getLocalDevice();
  agent = local.getDiscoveryAgent();
} catch (BluetoothStateException e) {
  throw new MIDletStateChangeException(
    "Unable to retrieve local Bluetooth device.");
}

deviceVector = new Vector();

addDevices();
try {
  agent.startInquiry(DiscoveryAgent.GIAC, this);
} catch (BluetoothStateException e) {
  throw new MIDletStateChangeException(
    "Unable to start the inquiry");
}

isInInquiry = true;

}
public void deviceDiscovered(RemoteDevice device,
  DeviceClass cod) {
  String address = device.getBluetoothAddress();
  deviceList.insert(0, address + "-I", null);

  deviceVector.insertElementAt(device, 0);

}
```

```
      public void inquiryCompleted(int type) {

          isInInquiry = false;

        Alert dialog = null;
        // Determine if an error occurred. If one did occur
        // display an Alert before allowing the
        // application to exit.
        if (type != DiscoveryListener.INQUIRY_COMPLETED) {
          dialog = new Alert("Bluetooth Error",
            "The inquiry failed to complete normally", null,
              AlertType.ERROR);
        } else {
          dialog = new Alert("Inquiry Completed",
            "The inquiry completed normally", null,
              AlertType.INFO);
        }
        dialog.setTimeout(Alert.FOREVER);
        Display.getDisplay(this).setCurrent(dialog);
      }
      private void addDevices() {
        // Retrieve the pre-known device array and append the
        // addresses of the Bluetooth devices. If there are no
        // pre-known devices, move on to cached devices.
        RemoteDevice[] list =
          agent.retrieveDevices(DiscoveryAgent.PREKNOWN);
        if (list != null) {
          for (int i = 0; i < list.length; i++) {
            String address = list[i].getBluetoothAddress();
            deviceList.insert(0, address + "-P", null);

            deviceVector.insertElementAt(list[i], 0);

          }
        }
```

```
      // Retrieve the cached device array and append the
      // addresses to the list.
      list = agent.retrieveDevices(DiscoveryAgent.CACHED);
      if (list != null) {
        for (int i = 0; i < list.length; i++) {
          String address = list[i].getBluetoothAddress();
          deviceList.insert(0, address + "-C", null);
```

```
          deviceVector.insertElementAt(list[i], 0);
```

```
        }
      }
    }
}
```

Now that the `DiscoveryMIDlet` keeps track of each `RemoteDevice` object found and when the MIDlet is performing an inquiry, the `DiscoveryMIDlet` can be modified to perform a service search. The service search is started when a user selects one of the devices displayed on the screen. Because many Bluetooth devices cannot start service searches while the device is performing an inquiry, the inquiry is canceled if one is occurring before the search is started.

```
public class DiscoveryMIDlet extends BluetoothMIDlet
  implements DiscoveryListener {

  ...
```

```
  /**
   * The List of service records that were found.
   */
  private List serviceRecordList;
```

```
  ...

  public void commandAction(Command c, Displayable d) {
```

```
if (c.getCommandType() == Command.EXIT) {

    if (isInInquiry) {

        // Try to cancel the inquiry.
        agent.cancelInquiry(this);

    }

    notifyDestroyed();

} else if (c == List.SELECT_COMMAND) {
    // Since the deviceList is currently visible, the user
    // must have selected a device to search so display
    // the serviceRecordList screen.
    serviceRecordList = new List("Services Found",
      List.IMPLICIT);
    serviceRecordList.addCommand(new Command("Exit",
      Command.EXIT, 1));
    serviceRecordList.setCommandListener(this);
    Alert splash = null;

    // If an inquiry is presently occurring, cancel the
    // inquiry before starting the service search.
    // Otherwise, start the service search
    if (isInInquiry) {
        agent.cancelInquiry(this);
        splash = new Alert("Cancel Inquiry", "Ending" +
          "the inquiry and starting the service search",
          null, AlertType.INFO);
    } else {
        splash = new Alert("Starting Search",
          "Starting the service search",
          null, AlertType.INFO);
        startServiceSearch();
```

```
      }
      splash.setTimeout(2000);
      Display.getDisplay(this).setCurrent(splash,
        serviceRecordList);
    }
  }
  /**
   * Starts the service search.
   */
  private void startServiceSearch() {
    try {
      // Search for the Bluetooth Game service record and
      // retrieve the name attribute in addition to the
      // default attributes.
      UUID[] uuidList = new UUID[1];
      uuidList[0] = new
        UUID("0FA1A7AC16A211D7854400B0D03D76EC", false);
      int[] attrList = new int[1];
      attrList[0] = 0x100;

      // The RemoteDevices are in the deviceVector in the
      // same order as they are on the screen so getting
      // the index allows us to retrieve the correct
      // RemoteDevice object.
      int index = deviceList.getSelectedIndex();
      RemoteDevice d =
        (RemoteDevice)deviceVector.elementAt(index);
      int id = agent.searchServices(attrList, uuidList,
        d, this);
    } catch (BluetoothStateException e) {
      Alert error = new Alert("Error",
        "Unable to start the service search (" +
        e.getMessage() + ")", null, AlertType.ERROR);
      error.setTimeout(Alert.FOREVER);
```

```
            Display.getDisplay(this).setCurrent(error,deviceList);
        }
    }
```

```
public void inquiryCompleted(int type) {
    isInInquiry = false;
    Alert dialog = null;

    // Determine if an error occurred. If one did occur
    // display an Alert before allowing the application
    // to exit.
    if (type != DiscoveryListener.INQUIRY_COMPLETED) {
```

```
        // If the device inquiry was terminated, then the
        // user must have selected a RemoteDevice to
        // perform a service search on so start the
        // service search.
        if (type == DiscoveryListener.INQUIRY_TERMINATED) {
            startServiceSearch();
            return;
        } else {
```

```
            dialog = new Alert("Bluetooth Error",
                "The inquiry failed to complete normally", null,
                AlertType.ERROR);
```

```
        }
```

```
    } else {
        dialog = new Alert("Inquiry Completed",
            "The inquiry completed normally", null,
            AlertType.INFO);
    }
```

```
      dialog.setTimeout(Alert.FOREVER);
      Display.getDisplay(this).setCurrent(dialog);
  }
}
```

The `startServiceSearch()` method is called from two different parts of the previous code. The `startServiceSearch()` method is called from the `commandAction()` method if an inquiry is not in progress. If an inquiry is in progress, the `commandAction()` method cancels the inquiry and the `startServiceSearch()` method is called from the `inquiryCompleted()` method when the cancel is processed.

The `startServiceSearch()` method starts the service search. This method performs a service search for the Bluetooth game service described earlier in this chapter. When a service is found with the Bluetooth game service's UUID, the 0x100 attribute also is retrieved. This is the ServiceName attribute ID as defined by the Bluetooth SIG. (The ServiceName is used in example code later in this chapter.)

Even though the `startServiceSearch()` method starts the search, the `DiscoveryMIDlet` does not do anything with the services that it finds at present. Therefore the `servicesDiscovered()` method is modified to display the number of service records returned to the `DiscoveryMIDlet`. The `serviceSearchCompleted()` method is also modified to display a message to the user when the service search ends.

```
public class DiscoveryMIDlet extends BluetoothMIDlet
  implements DiscoveryListener {

  ...
```

```
    public void servicesDiscovered(int transID,
      ServiceRecord[] record) {
      serviceRecordList.insert(0,
        Integer.toString(record.length), null);
    }
    /**
      * Called when the service search has ended. Displays a
      * message to the user that the service search
```

```
    * completed and specifies if the search completed
    * normally.
    *
    * @param transID the transaction ID
    * @param type specifies how the service search completed
    */
public void serviceSearchCompleted(int transID, int type) {
  Alert dialog = null;
  // Determine if an error occurred. If one did occur
  // display an Alert before allowing the application
  // to exit.
  if (type !=
    DiscoveryListener.SERVICE_SEARCH_COMPLETED) {
    dialog = new Alert("Bluetooth Error",
      "The service search failed to complete normally",
      null, AlertType.ERROR);
  } else {
    dialog = new Alert("Service Search Completed",
      "The service search completed normally", null,
      AlertType.INFO);
  }
  dialog.setTimeout(Alert.FOREVER);
  Display.getDisplay(this).setCurrent(dialog);
}

}
```

7.3.6 Working with Service Records

After a `ServiceRecord` is retrieved from a service search, the next step is to determine whether the service described by the `ServiceRecord` is the desired service. Once the service is determined to be the desired service, the `getConnectionURL()` method can be called to retrieve the connection string that establishes a connection to the service. This connection string may then be passed to `Connector.open()` to establish the connection. The `getConnectionURL()` method also allows the application to specify the security requirements of the connection and whether the local device is the master or the slave (see Section 4.3 for more information on Bluetooth security).

Before calling `getConnectionURL()`, the application must determine whether the `ServiceRecord` describes the service desired. This step highlights the need to be as specific as possible when determining the list of UUIDs to search for. Being as specific as possible minimizes the need to do additional work to determine whether the `ServiceRecord` returned is for the desired service. In most situations, a complete list of UUIDs used in the service search eliminates the need for additional verification. When the service record is discovered in this situation, all that is required is calling the `getConnectionURL()` method to begin using the service.

There are situations that require additional verification or determination. For example, if the local device is able to locate two instances of the same service, the local device could connect to the service that is currently less busy. The application may also want to request additional information that allows the user of the application to determine which service to use.

To actually access the values of each of the attributes, the `getAttributeValue()` method should be used. The `getAttribute-Value()` method returns the attribute value of the attribute ID specified or `null` if the attribute ID is not in this service record. The value of an attribute is encapsulated in the `DataElement` class. The `DataElement` class provides accessor methods to determine the type of the `DataElement` and its value. Table 7.14 lists the different types of `DataElements` and how these types relate to the `DataElement` class.

Before the value of a `DataElement` is retrieved, the `getData-Type()` method should be called to verify the data type of the value. This step should always be done before retrieving the value because calling the wrong method on a `DataElement` object causes a `Class-CastException` to be thrown.

To show how to use `DataElements` and `ServiceRecords`, we are modifying the `DiscoveryMIDlet` to display the ServiceName attribute value for each service record found. This is done by modifying the

Table 7.14   Bluetooth `DataElement` Types and Their Associated Java Types

Bluetooth Type	`DataElement` Data Type	Java Type	Method for Retrieving Value from `DataElement`
Null	NULL	Represents a `null` value	None
Unsigned integer (1 byte)	U_INT_1	`long` value in the range 0 to 255	`getLong()`
Unsigned integer (2 bytes)	U_INT_2	`long` value in the range 0 to $2^{16}-1$	`getLong()`
Unsigned integer (4 bytes)	U_INT_4	`long` value in the range 0 to $2^{32}-1$	`getLong()`
Unsigned integer (8 bytes)	U_INT_8	`byte[]` value in the range 0 to $2^{64}-1$	`getValue()`
Unsigned integer (16 bytes)	U_INT_16	`byte[]` value in the range 0 to $2^{128}-1$	`getValue()`
Integer (1 byte)	INT_1	`long` value in the range -128 to 127	`getLong()`
Integer (2 bytes)	INT_2	`long` value in the range -2^{15} to $2^{15}-1$	`getLong()`
Integer (4 bytes)	INT_4	`long` value in the range -2^{31} to $2^{31}-1$	`getLong()`
Integer (8 bytes)	INT_8	`byte[]` value in the range -2^{63} to $2^{63}-1$	`getValue()`
Integer (16 bytes)	INT_16	`byte[]` value in the range $-2^{127}-1$ to $2^{127}-1$	`getValue()`
URL	URL	`java.lang.String`	`getValue()`
UUID	UUID	`javax.bluetooth.UUID`	`getValue()`
Boolean	BOOL	`boolean`	`getBoolean()`
String	STRING	`java.lang.String`	`getValue()`
DataElement sequence	DATSEQ	`java.util.Enumeration`	`getValue()`
DataElement alternative	DATALT	`java.util.Enumeration`	`getValue()`

`servicesDiscovered()` method. First, the `servicesDiscovered()` method retrieves the `DataElement` for the ServiceName attribute. Before the string contained in the `DataElement` is extracted, the `get-DataType()` method must be called to determine the type of attribute value stored in the `DataElement` object. After it is verified that the `DataElement` is a `String`, the value of the `DataElement` is displayed.

```
public class DiscoveryMIDlet extends BluetoothMIDlet
   implements DiscoveryListener {

  ...
```

```
  /**
   * The service records that were found in the last
   * service search.
   */
  private Vector serviceRecordVector;
```

```
  ...
```

```
  /**
   * Called each time a service is discovered. Retrieve
   * the name attribute from the service record and
   * display it on the screen. Add the service
   * record to the serviceRecordVector for later.
   *
   * @param transID the transaction ID
   * @param record the service records that were found
   */
```

```
  public void servicesDiscovered(int transID,
    ServiceRecord[] record) {
```

```
    // Process each service record individually
    for (int i = 0; i < record.length; i++) {
      // Retrieve the name attribute from the service record
      DataElement nameElement =
        (DataElement)record[i].getAttributeValue(0x100);
      // The name attribute is only valid if it exists and
```

```
      // is a String. If either of these conditions fail,
      // move on to the next service record.
      if ((nameElement != null) &&
        (nameElement.getDataType() == DataElement.STRING)) {

        // Retrieve the name and display it on the screen.
        String name = (String)nameElement.getValue();
        serviceRecordList.insert(0, name, null);
        serviceRecordVector.insertElementAt(record[i], 0);
      }
    }
```

```
  }
  ...
  private void startServiceSearch() {
```

```
    serviceRecordVector = new Vector();
```

```
    try {
      // Search for the Bluetooth Game service record and
      // retrieve the name attribute in addition to the
      // default attributes.
      UUID[] uuidList = new UUID[1];
      uuidList[0] = new UUID(
        "0FA1A7AC16A211D7854400B0D03D76EC", false);
      int[] attrList = new int[1];
      attrList[0] = 0x100;

      // The RemoteDevices are in the deviceVector in the
      // same order as they are on the screen so getting
      // the index allows us to retrieve the correct
      // RemoteDevice object.
      int index = deviceList.getSelectedIndex();
      RemoteDevice d =
        (RemoteDevice)deviceVector.elementAt(index);
```

```
      int id = agent.searchServices(attrList, uuidList,d,
        this);
  } catch (BluetoothStateException e) {
    Alert error = new Alert("Error",
      "Unable to start the service search (" +
      e.getMessage() + ")", null, AlertType.ERROR);
    error.setTimeout(Alert.FOREVER);

    Display.getDisplay(this).setCurrent(error,deviceList);
  }
 }
}
```

In addition to displaying the service name of each service discovered, every ServiceRecord object returned to the application is stored in a Vector so that additional information can be gathered from the service record later.

7.3.7 Retrieving Additional Attributes after Service Discovery

The getAttributeIDs() method returns the IDs of all the attributes that have been retrieved from the remote device. This method does not return all the attributes defined in the service record on the remote device. Why would a ServiceRecord that has been discovered not have all the attributes of the service record on the remote device? The answer is simple. To reduce the amount of data actually sent over the air. If there is no intention to actually use an attribute, there is no reason to retrieve the attribute.

Sometimes the local device needs an attribute only in specific instances. JABWT provides a way to retrieve these attributes after the service search has been completed. For retrieving additional attributes, the populateRecord() method is called with the list of additional attributes to retrieve. The populateRecord() method returns true if some or all of the attributes specified are retrieved. The method may also throw an IOException if the remote device that has the service described by the ServiceRecord cannot be reached or the service is no longer available.

The `populateRecord()` method goes over the air to retrieve these additional attributes. Unlike `searchServices()`, which issues a request and then returns, a call to `populateRecord()` does not return until it fails or the information is retrieved. Because the `populateRecord()` method blocks, be aware of where this method is called. Calling the method within an event handler can affect the user's experience.

To show how to use the `populateRecord()` method, we modify the `DiscoveryMIDlet` to retrieve the ServiceDescription attribute. This procedure requires a few modifications to the `DiscoveryMIDlet`. Because it implements `Runnable` through the `BluetoothMIDlet` class, the `DiscoveryMIDlet` must have a `run()` method. When a user selects a service name from the `List` displayed on the screen, a new thread is created for the `DiscoveryMIDlet` that retrieves the Service-Description attribute by means of the `populateRecord()` method.

```
public class DiscoveryMIDlet extends BluetoothMIDlet
  implements DiscoveryListener {

  ...

  public void commandAction(Command c, Displayable d) {
    if (c.getCommandType() == Command.EXIT) {
      if (isInInquiry) {
        // Try to cancel the inquiry.
        agent.cancelInquiry(this);
      }
      notifyDestroyed();
    } else if (c == List.SELECT_COMMAND) {

      // Determine if the deviceList is the one that was
      // selected.
      if (d == deviceList) {

        // Since the deviceList is currently visible, the
        // user must have selected a device to search so
        // display the serviceRecordList screen.
        serviceRecordList = new List("Services Found",
```

```
      List.IMPLICIT);
serviceRecordList.addCommand(new Command("Exit",
  Command.EXIT, 1));
serviceRecordList.setCommandListener(this);
Alert splash = null;

// If an inquiry is presently occurring, cancel
// the inquiry before starting the service search.
// Otherwise, start the service search
if (isInInquiry) {
  agent.cancelInquiry(this);
  splash = new Alert("Cancel Inquiry",
    "Ending the inquiry and starting the service"
    + "search", null, AlertType.INFO);
} else {
  splash = new Alert("Starting Search",
    "Starting the service search",
    null, AlertType.INFO);
  startServiceSearch();
}

splash.setTimeout(2000);
Display.getDisplay(this).setCurrent(splash,
  serviceRecordList);
```

```
  } else {
    // Since the serviceRecordList is being
    // displayed, get the ServiceDescription
    // attribute for the service that was selected.
    new Thread(this).start();
  }
} else {
  // The user must have selected the Back command. So
  // display the names of all the services that were
  // found.
  serviceRecordList = new List("Services Found",
    List.IMPLICIT);
```

```
      serviceRecordList.addCommand(new Command("Exit",
        Command.EXIT, 1));
      serviceRecordList.setCommandListener(this);

      for (int i = 0; i < serviceRecordVector.size(); i++) {
        // Services were only added to the
        // serviceRecordVector if they had a valid name
        // attribute. Therefore, there is no need to test
        // the nameElement.
        ServiceRecord record =
          (ServiceRecord)serviceRecordVector.elementAt(i);
        DataElement nameElement =
          (DataElement)record.getAttributeValue(0x100);
        String name = (String)nameElement.getValue();
        serviceRecordList.insert(0, name, null);
      }

      Display.getDisplay(this).setCurrent(
        serviceRecordList);
    }
  }

  /**
   * This thread is started when the user selects a
   * service name. This thread retrieves the description
   * of the service and displays the description on the
   * screen.
   */
  public void run() {
    Alert error = null;
    try {
      // Identify the service record selected by the user
      int index = serviceRecordList.getSelectedIndex();
      ServiceRecord record = (ServiceRecord)
        serviceRecordVector.elementAt(index);
      // Retrieve the ServiceDescription attribute
```

```
      // from the remote device
      int[] attrList = new int[1];
      attrList[0] = 0x101;
      if (record.populateRecord(attrList)) {
        // Retrieve the ServiceDescription DataElement
        // and verify that it exists and is a String
        DataElement descriptionDataElement =
          record.getAttributeValue(0x101);

        if ((descriptionDataElement != null) &&
          (descriptionDataElement.getDataType() ==
          DataElement.STRING)) {

          // Display the description on the screen
          String description =
            (String)descriptionDataElement.getValue();
          Form descriptionForm = new Form(
            "Service Description");
          descriptionForm.append(description);
          descriptionForm.addCommand(new Command("Exit",
            Command.EXIT, 1));
          descriptionForm.addCommand(new Command("Back",
            Command.OK, 1));
          descriptionForm.setCommandListener(this);
          Display.getDisplay(this).setCurrent(
            descriptionForm);

          return;
        }
      }
      error = new Alert("Error",
        "Failed to retrieve the description of this service",
        null, AlertType.ERROR);
    } catch (IOException e) {
      error = new Alert("Error",
        "Failed to retrieve the description (IOException: "+
        e.getMessage() + ")", null, AlertType.ERROR);
```

```
    }
    // Display the error message on the screen
    error.setTimeout(2000);
    Display.getDisplay(this).setCurrent(error);
  }

}
```

Note that in the `run()` method, the ServiceDescription attribute is not simply retrieved and displayed. Like the ServiceName attribute displayed on the screen previously, the `DataElement` returned for the Service-Description attribute is inspected to determine whether the `DataElement` represents a `String`. Also, the return value of `populateRecord()` is checked to verify that the ServiceDescription attribute was retrieved.

In addition to starting the thread that retrieves the Service-Description attribute, the `commandAction()` method is modified to allow the user to return to the list of service names after retrieving the service description. This procedure allows the user to retrieve the Service-Description attribute of other services that were found during the service search.

7.3.8   Simple Device and Service Discovery

To make things easier for developers, the `selectService()` method combines the process of device and service discovery. The `select-Service()` method returns a connection string that can be used by `Connector.open()` to connect to the service. If a service cannot be found that meets the requirements of the search, `selectService()` returns `null`. The `selectService()` method has three arguments. The first argument is the UUID to search for in the ServiceClassIDList attribute. The second argument specifies the minimum security requirements needed for the connection. The third argument specifies whether the local device needs to be the master of this connection.

```
Connection getConnection(String uuidValue) throws
  IOException {
  String connString;
  try {
    // Retrieve the LocalDevice and DiscoveryAgent objects.
    LocalDevice local = LocalDevice.getLocalDevice();
    DiscoveryAgent agent = local.getDiscoveryAgent();

    // Retrieve the connection string to the service
    UUID searchUUID = new UUID(uuidValue, false);
    connString = agent.selectService(searchUUID,
      ServiceRecord.NOAUTHENTICATE_NOENCRYPT, false);
  } catch (BluetoothStateException e) {
    throw new IOException("BluetoothStateException: " +
      e.getMessage());
  }
  if (connString == null) {
    throw new IOException(
      "Failed to locate a device with the UUID " +
      uuidValue);
  }
  return Connector.open(connString);
}
```

The getConnection() method uses the selectService() method to locate a service using the UUID value specified. If a service is found, the code calls Connector.open() to establish a connection to the service found. If a service is not found, the code throws an IOException to signal that a connection could not be established.

JABWT does not specify how selectService() finds a device with the service requested. Therefore an inquiry can occur during the call to selectService(). Because of this possibility, the select-Service() method should be called in a separate thread because an inquiry can last 10 seconds or more. This procedure prevents the application from appearing frozen while the selectService() method is called.

Generating a UUID for Your Service

For testing your server application, you can make up an arbitrary series of hexadecimal digits for the UUID needed in the server's connection string. However, when you are ready to package your application, you should provide a UUID that is truly unique so that clients can use this UUID to locate your server application on Bluetooth devices. For example, the `defineGameService()` method shown earlier in this chapter used the UUID `0FA1A7AC16A211D7854400B0D03D76EC`. A process has been specified for generating UUIDs that have a very high probability of being unique [39]. This process generates a UUID using a timestamp and the Ethernet address of the computer used to generate the UUID. It would have been useful to provide a method for generating UUID as part of the JABWT. However, this UUID-generating method would have been used only during development and would not have been used for actually running JABWT applications. This method was omitted from the API to keep the JABWT implementation as small as possible for Java ME devices.

Several utilities are available for generating UUIDs. These utilities tend to be operating system specific because of the need to access the Ethernet card address. On Windows, UUIDs can be generated with the tool `Guidgen.exe` that comes with Microsoft® Visual Studio®. On Linux a `uuidgen` function is available as part of the e2fsprogs package for second extended (ext2) Linux file systems. For example, the command

```
uuidgen -t
```

returns a result such as the following:

```
0fa1a7ac-16a2-11d7-8544-00b0d03d76ec
```

The last twelve hexadecimal digits are the network card address. As can be seen from the example output from `uuidgen`, it is conventional to include hyphens between certain digits of a UUID. However, hyphens are not allowed in the string representation of JABWT UUIDs; the hyphens must be removed before the UUID can be used in a JABWT connection string.

The Bluetooth SIG has reserved a range of UUIDs for use by the Bluetooth specifications. This reserved range starts at the Bluetooth base value of `00000000-0000-1000-8000-00805F9B34FB` and includes all 2^{32} values up to `FFFFFFFF-0000-1000-8000-00805F9B34FB`.

Applications should use values in this range only for the purposes described in the Bluetooth specifications.

7.4 Summary

A Bluetooth client application communicates with a Bluetooth server application on another device to use the services provided by the server. Service discovery is the process of identifying the services offered by nearby Bluetooth devices by retrieving their service records. This chapter describes the JABWT methods used for locating service records that contain a particular collection of UUIDs and for retrieving some of the attributes of those service records.

Service registration is the process of creating a service record describing a service and adding it to the SDDB, where it can be discovered by clients. For many server applications, developers do not need to be concerned about service records. The JABWT implementation creates and registers a service record automatically. This service record advertises the service to potential clients and provides the information needed to construct a connection string that clients can use to access the service. In many cases, this automatically generated service record is sufficient.

In cases in which the automatically generated service record is inadequate, JABWT provides capabilities that allow developers to modify the service records. This chapter describes the service records automatically generated by JABWT and the procedure for modifying those service records when necessary.

8 L2CAP

CHAPTER

This chapter covers the following topics:

- What is L2CAP?
- What kinds of applications are appropriate for L2CAP communications?
- What support for L2CAP communication does JABWT provide?
- How is an L2CAP channel configured by a JABWT application?
- Why do L2CAP applications need to provide flow control?

8.1 Overview

L2CAP stands for *logical link control and adaptation protocol.* L2CAP is a multiplexing layer that allows several higher-level protocols or applications to use Bluetooth communications. Figure 8.1 shows where L2CAP fits in the Bluetooth protocol stack. The figure shows a common hardware configuration with separate host hardware and a Bluetooth radio module. In this configuration, the L2CAP layer is on the host side of the HCI. Also in this configuration, L2CAP uses HCI to communicate with the baseband layer in the Bluetooth radio module. All Bluetooth data communications use L2CAP, but Bluetooth voice communications do not.

As Figure 8.1 shows, several protocols sit on top of the L2CAP layer and use L2CAP to provide access to the Bluetooth hardware in the Bluetooth radio module. The SDP and the RFCOMM protocol are two of these higher-level protocols. JABWT provides access to both SDP and RFCOMM. Chapter 7 describes how JABWT applications use the SDP protocol to discover service records. Chapter 4 describes how JABWT applications use the RFCOMM protocol for serial port communications.

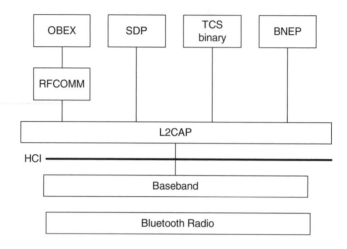

Figure 8.1   Position of L2CAP in the Bluetooth stack.

JABWT does not provide access to the other two protocols shown above L2CAP in Figure 8.1. TCS binary defines the call-control signaling that establishes speech and data calls between Bluetooth devices. The BNEP [5] is an optional protocol developed after Bluetooth specification version 1.1 but based on the 1.1 version of the specification. The BNEP can be used to transmit IP packets over L2CAP and supports the PAN profile [40].

The PAN profile is one of three Bluetooth profiles that provide access to the Internet for Bluetooth devices. Two other profiles for Internet access are the LAN Access Profile and the Dial-up Networking Profile. The LAN Access Profile and the Dial-up Networking Profile use the RFCOMM protocol rather than the BNEP as their entry point into the Bluetooth protocol stack. The LAN Access Profile addresses the case in which a data terminal such as a laptop uses Bluetooth wireless technology to communicate with a LAN access point that serves as a gateway to a LAN. This is one of the use cases also addressed by the PAN profile. The PAN profile supersedes the LAN Access Profile, so the LAN Access Profile is now considered obsolete. The Dial-up Networking Profile describes the case in which a data terminal such as a laptop uses Bluetooth wireless technology to communicate with a cell phone or modem that provides a dial-up connection to a LAN.

8.1.1   L2CAP Channels and L2CAP Packets

Figure 8.2 illustrates the multiplexing service that L2CAP provides. The left side of Figure 8.2 represents one Bluetooth device, and the right side represents another Bluetooth device. On the left side, an L2CAP server application and the RFCOMM protocol both are using L2CAP to provide Bluetooth communications. L2CAP establishes L2CAP channels that connect these higher-level entities to their counterparts on the remote device. In Figure 8.2, two L2CAP channels are represented. One L2CAP channel runs between the L2CAP server application and the L2CAP client application. Another L2CAP channel runs between the RFCOMM protocol layers on the two devices. In Figure 8.2 these L2CAP channels are represented as highways. The arrows traveling over the highways represent L2CAP packets. L2CAP provides full-duplex communications, so the L2CAP packets can travel in both directions.

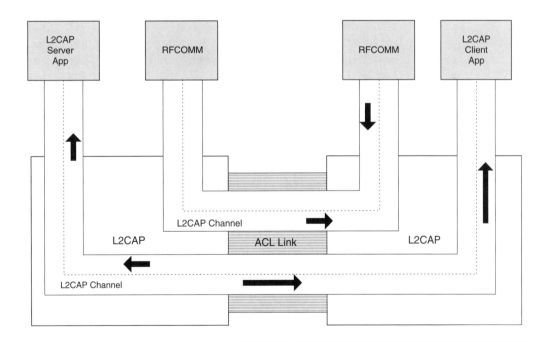

Figure 8.2   L2CAP channels transmit L2CAP packets to multiple destinations.

The length of the arrow represents the size of the L2CAP packet. There is a limit to the size of the L2CAP packet that can be transferred over a particular L2CAP channel in a particular direction. This limit is called the *maximum transmission unit* (MTU). In Figure 8.2, the arrows heading to the L2CAP client are longer than the arrows heading from the L2CAP client. This configuration is intended to suggest that the MTU for the L2CAP packets received by the L2CAP client is larger than the MTU for packets received by the L2CAP server. When the L2CAP channel is established, the MTU values for travel in both directions are negotiated by the two L2CAP components. Section 8.2.3 provides additional details about MTUs.

Figure 8.2 shows the ACL link between the two devices. The baseband layer of the protocol stack establishes the ACL link. There is exactly one ACL link between two Bluetooth devices communicating with each other. The ACL link and the baseband functions provide the infrastructure needed to support the high-level, "logical" abstractions of L2CAP channels and L2CAP packets that L2CAP presents to higher-level protocols and applications.

L2CAP packets have to be converted into one or more baseband packets for transmission over the ACL link. The receiving device then reassembles these baseband packets into L2CAP packets. There are various sizes of baseband packets, but the largest payload is 339 bytes. This is much smaller than the largest payload possible for an L2CAP packet, 65,535 bytes. Because baseband packets are much smaller than packet sizes used by higher-level protocols and applications, the segmentation and reassembly process hides the details about Bluetooth baseband packets from the higher levels of the stack and from applications. By presenting abstractions such as L2CAP channels and L2CAP packets to higher levels, L2CAP makes it easier for higher-level protocols and applications to use Bluetooth communications. This adaptation function is one of the important contributions of L2CAP.

The L2CAP channels shown in Figure 8.2 are what are known as *connection-oriented channels*. They support L2CAP packet transmission in both directions, and the L2CAP packets transmitted are intended for use only by the single device at the other end of the ACL link. L2CAP also provides *connectionless channels*. Connectionless channels allow only one-way traffic, and they are intended for broadcasting L2CAP packets to a group of nearby devices. JABWT provides no support for connectionless channels.

8.1.2 Reasons for Using L2CAP

A number of Bluetooth protocols are defined on top of the L2CAP protocol. Figure 8.1 shows four of these protocols: RFCOMM, SDP, TCS binary, and BNEP. However, the number of these protocols keeps growing as new Bluetooth profiles are developed that define new protocols based on L2CAP. Because JABWT does not provide an interface to all of the L2CAP features, it may not be possible to implement all of the protocols and profiles defined by the Bluetooth SIG with JABWT. For example, the Telephony Control Protocol (TCS-BIN) makes use of both connection-oriented and connectionless L2CAP channels. Because JABWT does not support connectionless L2CAP channels, it would not be possible to implement all of the TCS-BIN protocol.

In the case of TCS-BIN, the JSR-82 expert group made an explicit decision that telephony control would likely be provided by device manufacturers, so it was not necessary for JABWT to provide an API that could be used to implement TCS-BIN. Consequently, it may or may not be possible to use JABWT to implement all of the protocols and profiles defined by the Bluetooth SIG. Developers have to study the specifications carefully and assure themselves that JABWT support for L2CAP is sufficient to implement a particular protocol or profile.

For example, JABWT and the Bluetooth audio/video profiles have incompatible requirements regarding flush timeout. The baseband layer, a layer below L2CAP in the Bluetooth stack (Figure 8.1), offers the option of retransmitting packets until they are received successfully. The retransmit option is controlled by a parameter called *flush timeout,* which indicates how long the baseband attempts to retransmit a packet before giving up and flushing the packet. JABWT currently requires that L2CAP communications use a flush timeout of 0xFFFF, which means that baseband should never give up. The baseband should continue to retransmit until either the packet is acknowledged or the ACL link terminates. On the other hand, the Bluetooth audio/video profiles specify that applications using the audio/video profiles be allowed to set the value of flush timeout. These audio/video profiles recommend that small values should be used for flush timeout. Small values for flush timeout help ensure that most of the L2CAP bandwidth is devoted to the initial transmission of audio/video data and that retransmission is minimized.

The next revision of the JABWT specification is expected to remove the requirement for using a flush timeout of 0xFFFF. The revised JABWT specification could introduce a way for applications to request a particular flush timeout. This would remove the current obstacle to using JABWT to implement the audio/video profiles.

In addition to the standardized applications defined as Bluetooth profiles, custom applications based on the L2CAP API are possible. In the case of custom applications, JABWT developers have a choice about whether to use the L2CAP API, the RFCOMM API, or the OBEX API. For most custom applications, the stream-oriented APIs provided by the higher-level protocols RFCOMM and OBEX would have advantages. However, if a custom application requires control over which bytes are sent together in a single packet, then the L2CAP API is the best choice. This might be the case if the application needs to define a new packet-based protocol analogous to BNEP or SDP. If a stream-based protocol can be used, RFCOMM and OBEX are better options. OBEX itself is a good example of a stream-based protocol that can be defined over RFCOMM.

8.2   API Capabilities

This section describes the support that JABWT provides for L2CAP communications. The Java interfaces defined by JABWT for L2CAP are described. Examples of connection strings used to open an L2CAP connection are provided. Guidelines are proposed for configuring L2CAP channels with MTU parameters in connection strings and for selecting appropriate sizes for the byte arrays used to send and receive L2CAP packets. Issues related to flow control are discussed.

8.2.1   JABWT for L2CAP

Figure 8.3 shows the two interfaces defined in JABWT for L2CAP communications, L2CAPConnection and L2CAPConnectionNotifier. An L2CAP server uses an L2CAPConnectionNotifier to wait for an L2CAP client to establish a connection. The notifier then returns an L2CAP-Connection object to provide access to the L2CAP channel between the client and the server. The L2CAPConnection interface can be used to send data between the client and the server using the L2CAP protocol.

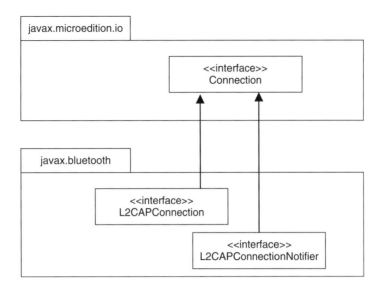

Figure 8.3 JABWT defines two interfaces for L2CAP communications.

When an L2CAP client is successful in opening a connection to an L2CAP server, the result returned from Connector.open() is an L2CAPConnection that gives the client access to the L2CAP channel. Again, the client uses the L2CAPConnection to send and receive data. The L2CAP client does not use the L2CAPConnectionNotifier interface.

For serial port (RFCOMM) communications, the JABWT specification did not need to define any new classes or interfaces beyond those in the GCF defined by CLDC. GCF already provided the stream-oriented interfaces StreamConnection and StreamConnectionNotifier. However, these CLDC interfaces were not useful for L2CAP because L2CAP communications are based on packets not on streams.

CLDC has a DatagramConnection interface that is packet oriented rather than stream oriented, and DatagramConnection was considered for possible use with L2CAP. As it turns out, send() and receive() methods were defined for L2CAPConnection. DatagramConnection has these same methods, but the method arguments are different. For DatagramConnection these methods take a Datagram argument, whereas the arguments to the L2CAPConnection versions of the methods are byte arrays. Each `Datagram` object sent over a `DatagramConnection` contains a destination address, because each `Datagram` can be sent to a different recipient. However, this overhead associated with `Datagram` is not needed for L2CAP packets. Each `L2CAPConnection` instance defines a unique sending application and a unique receiving application. Because the address information is contained in the connection, there is no need to provide an address as part of the argument to `send()` or `receive()`.

The `receive()` method blocks until either an L2CAP packet is read or the L2CAP channel is closed. The `ready()` method defined by the `L2CAPConnection` interface makes it possible to check whether a call to `receive()` will block. The `ready()` method returns `true` if an L2CAP packet is available to be read immediately by `receive()` without blocking.

8.2.2   Open an L2CAP Connection

Here are some examples of legal arguments to `Connector.open()` for L2CAP clients:

```
"btl2cap://0050CD00321B:1001"
"btl2cap://0050CD00321B:1001;receiveMTU=512"
"btl2cap://0050CD00321B:1001;receiveMTU=512;"
  + "transmitMTU=512"
"btl2cap://0050CD00321B:1001;authenticate=true;"
  + "encrypt=true"
"btl2cap://0050CD00321B:1001;master=true"
```

The protocol name `btl2cap` is "bee tee *el* two cap," not "bee tee *one* two cap." The entry `0050CD00321B` in these examples is the Bluetooth address of the server device. The entry `1001` in these examples is the PSM value for the server application. The PSM is obtained from the service record. The PSM tells the L2CAP layer on the remote device which server application the client wants as the destination of the new L2CAP channel. Higher-level protocols such as RFCOMM and SDP have L2CAP PSM values permanently assigned to them. A PSM has to be dynamically generated for an application, and it is possible that a different PSM will be assigned to the same server application the next time it starts up. This situation is similar to that of `btspp` server channel identifiers, which can vary from device to device and can even vary over time for the same service. One difference between PSMs and service channel identifiers in JABWT is that PSMs in `btl2cap` connection strings are interpreted as hexadecimal numbers, whereas service channel identifiers in `btspp` and `btgoep` connection strings are interpreted as decimal numbers.

The security parameters `authenticate` and `encrypt` are familiar from Chapter 4. The parameter `master` used to request the master role in the resulting Bluetooth network also is familiar. The parameters `receiveMTU` and `transmitMTU` are unique to L2CAP. The parameter `receiveMTU` indicates the size in bytes of the payload of the largest L2CAP packet that the client is willing to receive from the server. The parameter `transmitMTU` indicates the size in bytes of the payload of the largest L2CAP packet that the client will send to the server. These parameters are discussed further in the next section.

L2CAP client applications typically get a connection string from one of the instance methods:

- `ServiceRecord.getConnectionURL(int requiredSecurity, boolean master)`
- `DiscoveryAgent.selectService(UUID uuid, int security, boolean master)`

However, if either of the parameters `transmitMTU` or `receiveMTU` is to be used, it must be appended to the Strings returned from these methods. There is no option to include these MTU parameters in the arguments to the methods as there is for the security parameters and the master parameter.

Here are some examples of legal arguments to `Connector.open()` for L2CAP servers:

```
"btl2cap://localhost:9C68A2AA1EC011D79E6C00B0D03D76EC"
"btl2cap://localhost:9C68A2AA1EC011D79E6C00B0D03D76EC;"
                              + "name=L2CAPEx"
"btl2cap://localhost:9C68A2AA1EC011D79E6C00B0D03D76EC;"
                              + "receiveMTU=512"
"btl2cap://localhost:9C68A2AA1EC011D79E6C00B0D03D76EC;"
                              + "receiveMTU=512;"
                              + "transmitMTU=1024"
```

```
"btl2cap://localhost:9C68A2AA1EC011D79E6C00B0D03D76EC;"
                            + "authenticate=true;"
                            + "encrypt=true;"
                            + "authorize=true"
"btl2cap://localhost:9C68A2AA1EC011D79E6C00B0D03D76EC;"
                            + "master=true"
```

The only new parameters for L2CAP servers are `receiveMTU` and `transmitMTU`. They have the same meaning for the server as they do for the client. A PSM value is not part of the server's connection string. The PSM is generated automatically and inserted into the service record by the JABWT implementation. This is similar to the situation for `btspp` servers, in which the server channel identifier also is automatically generated and inserted into the service record (see Chapter 7).

8.2.3   L2CAP Channel Configuration

There may be a limit on the size of L2CAP payload the Bluetooth stack can receive or the size of payload a JABWT application is prepared to receive. The largest L2CAP packet payload the Bluetooth stack can receive is returned by `LocalDevice.getProperty("bluetooth.l2cap.receiveMTU.max")`.

The application should decide whether it wants to handle payloads as large as the stack can handle or something smaller. The answer is likely to depend on the nature of the application and the Java heap space expected to be available to applications.

On the basis of these considerations, the application can communicate `receiveMTU` to the remote device when it creates a connection using the `receiveMTU` parameter:

```
Connector.open("btl2cap://...;receiveMTU=1024")
```

For some applications it is important to be able to send L2CAP packets up to a particular size. For example, the BNEP needs to be able to transmit the maximum Ethernet packet payload, 1500 bytes, plus all of the associated BNEP headers in a single L2CAP packet. Consequently, BNEP needs to be able to send L2CAP packets with payloads of at least 1691 bytes. This requirement on the size of outgoing L2CAP packets can be declared by a JABWT application with the connection string parameter `transmitMTU=1691`.

In general, it is better not to specify `receiveMTU` or `transmitMTU` values in the connection string unless absolutely necessary. MTU values are assigned automatically if no MTU values are mentioned in the connection string. The automatic assignment usually is the default MTU of 672 defined by the L2CAP specification. However, there are cases, such as the BNEP case discussed earlier, in which setting a particular value is required.

The documentation on MTUs can be confusing. Many of the details about `receiveMTU` and `transmitMTU` in the JABWT specification are relevant only to implementers of the JABWT specification. The Bluetooth L2CAP specification [1] describes MTUs from the point of L2CAP. However, L2CAP allows a back-and-forth negotiation process for the MTU values for an L2CAP connection that does not really apply to JABWT applications. The `receiveMTU` and `transmitMTU` values specified in connection strings by JABWT applications should not be viewed as initial proposals in a back-and-forth negotiation. Instead they should be viewed as non-negotiable requirements.

For application developers, we boil down the essentials about MTUs to four rules for code development and one potential pitfall to be aware of even if all four rules are followed.

MTU Rule 1

The values for `receiveMTU` and `transmitMTU` in your L2CAP connection string must be no smaller than `L2CAPConnection.MINIMUM_MTU`, which is 48, the minimum MTU allowed by the Bluetooth L2CAP specification. The values for `receiveMTU` and `transmitMTU` also must be no larger than 65,535, the maximum payload size in an L2CAP packet.

MTU Rule 2

The value for `receiveMTU` in your L2CAP connection string must be smaller than or equal to `LocalDevice.getProperty("bluetooth.l2cap.receiveMTU.max")`, the largest L2CAP packet that can be received by the Bluetooth stack on the device on which your application is currently running. Applications can use this property at runtime to tailor their MTU values to the limits of any Bluetooth stack in use.

MTU Rule 3

For transmitting outgoing packets over an L2CAP connection with `send(byte[] outBuf)`, the byte array `outBuf` must be no larger than `L2CAPConnection.getTransmitMTU()`. If `outBuf` is larger than this, bytes are discarded before the L2CAP packet is sent. If `transmitMTU` was declared in the L2CAP connection string, then `getTransmitMTU()` has that same value.

MTU Rule 4

To receive incoming packets with `receive(byte[] inBuf)`, allocate a byte array of size `L2CAPConnection.getReceiveMTU()`.

If you use an `inBuf` smaller than `L2CAPConnection.get-ReceiveMTU()`, any bytes received in the L2CAP packet that do not fit in `inBuf` are discarded. If `receiveMTU` was declared in the L2CAP connection string, you could allocate a byte array of size `receiveMTU`. However, if the remote device declares a `transmitMTU` in its connection string that is smaller than `receiveMTU`, then `L2CAPConnection.get-ReceiveMTU()` could be smaller than `receiveMTU`. It never is larger than `receiveMTU`.

If you follow all four MTU rules, it is still possible that at runtime a particular L2CAP client and server will be unable to form a connection because of incompatible MTU values. For example, suppose Application A specifies MTU values in its connection string as follows:

```
"btl2cap://...;receiveMTU=receiveMTU_A;"+
   "transmitMTU=transmitMTU_A"
```

Also, suppose Application B specifies MTU values in its connection string as follows:

```
"btl2cap://...;receiveMTU=receiveMTU_B;"+
   "transmitMTU=transmitMTU_B"
```

MTU Mismatch Pitfall

The applications fail to connect because of inappropriate MTU values:

- If the largest packet Application A will send, $transmitMTU_A$, is larger than the largest packet Application B can receive, $receiveMTU_B$, or

- If the largest packet Application B will send, $\mathtt{transmitMTU_B}$, is larger than the largest packet Application A can receive, $\mathtt{receiveMTU_A}$.

The basic problem that leads to the MTU mismatch pitfall is that the L2CAP protocol does not provide any way to inquire at runtime about the MTU requirements of the remote device other than trying to make a connection and seeing whether you succeed. (Server applications can use custom service attributes in their service records to communicate their MTU requirements, but we are not aware of any precedent for doing this in the Bluetooth profiles.)

If you are writing both the client and server applications, you can avoid this pitfall by not specifying MTU values at all in the connection string or by making $\mathtt{transmitMTU_A} = \mathtt{receiveMTU_B}$ and $\mathtt{receiveMTU_A} = \mathtt{transmitMTU_B}$. If you have to interoperate with a variety of implementations, and they use different MTU values, the best strategy is to omit the $\mathtt{transmitMTU}$ parameter from your connection string. By omitting a $\mathtt{transmitMTU}$, you avoid a mismatch with the $\mathtt{receiveMTU}$ of the remote device.

Omitting a $\mathtt{receiveMTU}$ in your connection string does not provide the same benefit. The L2CAP channel configuration process requires that each application propose an MTU value for incoming L2CAP packets. If there is no $\mathtt{receiveMTU}$ in the connection string, then the JABWT implementation supplies a value for $\mathtt{receiveMTU}$ by using the constant $\mathtt{L2CAPConnection.DEFAULT\_MTU}$, which has the value 672.

8.2.4 No Flow Control in L2CAP

In Bluetooth specification version 1.1, there was no L2CAP flow control. The baseband layer, which is below L2CAP in the Bluetooth stack, provides flow control for the ACL link as a whole. Unfortunately, it is not sufficient to rely on the flow control provided by the lower baseband level. The problem is that L2CAP is a multiplexing layer that provides multiple L2CAP channels headed to multiple higher-level protocols or applications. In cases in which L2CAP packets arrive faster than they can be processed by one of the higher-level protocols and applications, the

L2CAP buffers fill up. When buffers are going to overflow, the only options available to L2CAP are the following:

- Let the lower-level flow control mechanisms kick in and shut off all incoming packets over this ACL link. Baseband flow control shuts off L2CAP packets over all the L2CAP channels, not just the L2CAP channels that are having trouble keeping up.
- Discard some L2CAP packets because there is no room for them in the L2CAP buffers. The lower stack layers will not retransmit these packets, because they have already been acknowledged as successfully received.

Simply discarding L2CAP packets is an unattractive option that would lead to data corruption or hung communications. However, the other option whereby the lower-level flow control shuts off all of the L2CAP channels can lead to deadlock in certain situations. The same problem of a multiplexing layer over a reliable communication layer arises in infrared data communications. The deadlock scenario for the IrDA protocol stack is discussed by Williams and Millar [46].

A summary of the deadlock scenario in L2CAP terms is as follows: Suppose a higher-level application uses two L2CAP channels. One of the L2CAP channels is used as a data channel, and the other L2CAP channel is used as a signaling channel. It is possible that the L2CAP buffers are overflowing because the application is waiting to receive an L2CAP packet on the signaling channel before it processes the packets on the data channel. If this is the case, and if baseband flow control shuts off the entire ACL link, then the L2CAP packet on the signaling channel cannot get through. The application continues to wait for this packet, so it does not process the packets on the data channel. However, processing packets on the data channel is the only thing that will free L2CAP buffers and get packets flowing again on the ACL link.

Because the original L2CAP specification had no flow control, the protocols and profiles that use L2CAP communications need to have their own mechanisms for flow control. For example, the RFCOMM protocol offers a credit-based flow control mechanism. The Hardcopy Cable Replacement Profile [6], which is one of three Bluetooth profiles for printing, also uses a credit-based flow control mechanism. The Bluetooth Extended Service Discovery Profile [41] uses an end-to-end window flow control mechanism.

A flow control mechanism, which operates on each L2CAP channel, has been defined for L2CAP in Bluetooth specification version 1.2. However, this flow control is an optional part of the L2CAP specification, so there is no assurance that every Bluetooth stack will implement this new flow control mechanism.

8.2.5   Types of Applications Using L2CAP

L2CAP applications can be implementations of standard Bluetooth profiles, or they can be nonstandardized, custom applications. The issues that come up are somewhat different depending on which kind of application is planned.

Implementing Bluetooth Profiles Using L2CAP

Certain Bluetooth profiles use L2CAP as their entry point into the Bluetooth protocol stack. Developers who intend to implement one of these Bluetooth profiles with JABWT applications need to study the profile specifications closely so that their applications can pass any Bluetooth qualification tests for this profile and can successfully interoperate with other devices that support these profiles.

Some Bluetooth profiles place requirements on MTU values configured for L2CAP channels. For example, the Hardcopy Cable Replacement Profile establishes two L2CAP channels: a control channel and a data channel. The profile requires that the MTU for the control channel be at least 128 bytes in both directions. The profile recommends that the MTU for the data channel be larger than the minimum (48 bytes) but does not require a particular value.

Implementing Custom Applications Using L2CAP

L2CAP applications that do not claim to conform to a Bluetooth profile do not need to undergo the Bluetooth qualification process. However, developers planning a custom application will still benefit from studying the Bluetooth profiles that use L2CAP. These profiles provide useful examples of how to best use L2CAP.

You should consider how your application will provide flow control. Several flow control schemes have been adopted by the Bluetooth profiles and protocols. For example, SDP entails a simple scheme that requires that only one SDP request from an SDP client to an SDP server can be outstanding at any point in time. Until the server responds to this request, the client is not allowed to issue another request over this same L2CAP channel.

Other current approaches for flow control over L2CAP are referenced in Section 8.2.4. It is worth consulting those references to see what options are available.

8.3   Programming with the API

This section shows example code for MIDP applications that use JABWT L2CAP communications. The example code illustrates the use of JABWT and some design considerations for L2CAP applications. Not all of the code needed to produce a running application is presented here. The complete code is available on this book's Web site at www.mkp.com.

8.3.1   Example: Echo L2CAP Packets

The example code shows both an L2CAP server and an L2CAP client. The L2CAP server echoes back any L2CAP packets sent by the L2CAP client. The client sends 50,000 bytes in a series of L2CAP packets. The size of the packets sent by the client is determined by the value of `getTransmitMTU()` for the connection. The payload of every packet the server receives is immediately sent back to the client. Both the client and the server keep a count of the total number of bytes sent or received over this connection. This byte count is reported when the packet exchange is complete.

We first look at the MIDlet method `openL2CAPConnection()`, which computes the connection string for either the server or the client and then starts it running. This method takes as arguments `receiveMTU` and `transmitMTU` and two boolean arguments that indicate whether those arguments should be added as parameters to the connection string. There is also an argument indicating whether a client or a server should be started. In the case in which a client should be started, `open-L2CAPConnection()` uses the `selectService()` method to obtain a connection string for a server application. The `selectService()` method attempts to find a server application that uses a particular UUID in its service record. Having computed the connection string in the `url` variable, the `openL2CAPConnection()` method starts a thread to execute the client. In the case in which the method is starting a server, computing the connection string is just a matter of concatenating strings for the `btl2cap` scheme, a UUID for the ServiceClassID (see Chapter 7), and any MTU parameters.

```
void openL2CAPConnection(boolean isClient,
  boolean receiveMTUInput,
  int receiveMTU,
  boolean transmitMTUInput,
  int transmitMTU) {
  String url = null;
  String paramString = "";
  if (receiveMTUInput) {
    paramString += ";receiveMTU=" + receiveMTU;
  }

  if (transmitMTUInput) {
    paramString += ";transmitMTU=" + transmitMTU;
  }

  if (isClient) {
    displayField.setText(
      "searching, please wait...");
    DiscoveryAgent agent =
      device.getDiscoveryAgent();

    try {
      url = agent.selectService(uuid,
        ServiceRecord.NOAUTHENTICATE_NOENCRYPT,
        false);
    } catch (BluetoothStateException e) {
      displayError("Error",
        "BluetoothStateException: " + e.getMessage());
    }
```

```
      if (url == null) {
        displayError("Error",
          "failed to find server!");
        return;
      }
      url += paramString;
      new  L2capClient(this).start(url);
    } else {
      url = "btl2cap://localhost:" + uuid.toString();
      url += paramString;
      new L2capServer(this).start(url);
    }
}
```

The `L2capServer` runs in its own thread, and it establishes an `L2CAPConnection` in its `run()` method.

```
public class L2capServer extends EchoParticipant
  implements Runnable {
  L2capMtuMIDlet parent;
  private String url;
  public L2capServer(L2capMtuMIDlet parent) {
    this.parent = parent;
    this.out = parent.displayField;
  }
  public void start(String url) {
    this.url = url;
    new Thread(this).start();
  }
  public void run() {
    LocalDevice device = null;
    L2CAPConnectionNotifier notifier = null;
    try {
      device = LocalDevice.getLocalDevice();
      /* Request that the device be made discoverable */
      device.setDiscoverable(DiscoveryAgent.GIAC);
    } catch(BluetoothStateException e) {
```

```
      parent.displayError("Error",
        "BluetoothStateException: " +
        e.toString());
      return;
    }
    try {
      notifier = (L2CAPConnectionNotifier)
        Connector.open(url);
    } catch (IllegalArgumentException e) {
      parent.displayError("Error",
        "IllegalArgumentException in" +
        "Connector.open()");
    } catch (IOException e) {
      parent.displayError("Error",
        "IOException: " +
        e.getMessage());
    }
    if (notifier == null) {
      return;
    }
    try {
      out.setLabel("["+url+"]");
      for (;;) {
        L2CAPConnection conn = notifier.acceptAndOpen();
        echoReceivedL2capPackets(conn);
        conn.close();
      }
    } catch(IOException e) {
      parent.displayError("Error",
        "IOException: " +
        e.getMessage());
    } catch (IllegalArgumentException e) {
      parent.displayError("Error",
        "IllegalArgumentException: " +
        e.getMessage());
    }
  }
}
```

The statement that creates the `L2CAPConnectionNotifier` is

```
notifier = (L2CAPConnectionNotifier)Connector.open(url);
```

This statement also creates an L2CAP service record (see Chapter 7). The `openL2CAPConnection()` method described above provides the `url`. Its value in this first example is

```
"btl2cap://localhost:9C68A2AA1EC011D79E6C00B0D03D76EC;"
                        + "receiveMTU=672;transmitMTU=672"
```

If the `url` argument to `Connector.open()` violates either MTU Rule 1 or MTU Rule 2, then this statement throws an `IllegalArgument-Exception`. There is code for exception handling in `L2capServer` to catch the `IllegalArgumentException`. It is unusual to provide an exception handler for an unchecked Java exception such as this, but it has advantages for demonstrating the results of MTU rule violations as described in the next section.

The statement that adds the service record to the SDDB and waits for a client to connect is

```
L2CAPConnection conn = notifier.acceptAndOpen();
```

If this statement does not throw an exception, it returns an instance of an `L2CAPConnection`, which provides access to the L2CAP channel between the `L2capServer` and the `L2capClient`.

The definition of the `echoReceivedL2capPackets()` method that actually sends and receives the bytes is described below in the `EchoParticipant` class. The `L2capServer` extends the `Echo-Participant` class, so it inherits this method.

The code for the `L2capClient` is shown next. Here the key statement in the `run()` method is

```
conn = (L2CAPConnection)Connector.open(url);
```

Again, the `openL2CAPConnection()` method provides the `url`; its value varies with the Bluetooth device address of the server and the PSM assigned to the server application. In this first example it has the value

```
"btl2cap://0050CD00321B:1001;authenticate=false;"
  + "encrypt=false;master=false;"
  + "receiveMTU=672;transmitMTU=672"
```

The Connector.open(url) statement attempts to form an L2CAP connection to the echo service described by the url argument.

```
public class L2capClient extends EchoParticipant
  implements Runnable {
  L2capMtuMIDlet parent;
  private String url;
  public L2capClient(L2capMtuMIDlet parent) {
    this.parent = parent;
    this.out = parent.displayField;
  }
  public void start(String url) {
    this.url = url;
    new Thread(this).start();
  }
  public void run() {
    L2CAPConnection conn = null;
    out.setLabel("["+url+"]");
    try {
      conn = (L2CAPConnection)Connector.open(url);
    } catch (IllegalArgumentException e) {
      parent.displayError("Error",
        "IllegalArgumentException in "
        +  "Connector.open()\n"
        +  e.getMessage());
    } catch (BluetoothConnectionException e) {
      String problem = "";
      if (e.getStatus() ==
        BluetoothConnectionException.UNACCEPTABLE_PARAMS)
      {problem = "unacceptable parameters\n"; }
      parent.displayError("Error",
        "BluetoothConnectionException: "
        + problem + "msg=" +
        e.getMessage() +
        "\nstatus= " + e.getStatus());
    } catch (IOException  e) {
```

```
      parent.displayError("Error",
        "IOException: " + e.getMessage());
    }

    if (conn == null) {
      return;
    }

    try {
      sendL2capPacketsForEcho(conn);
      conn.close();
    } catch (IOException e) {
      parent.displayError("Error",
          "IOException: " + e.getMessage());
    }
  }
}
```

If a connection cannot be formed between the L2CAP client and the L2CAP server because of incompatible MTU values (see the MTU mismatch pitfall in Section 8.2.3) then a BluetoothConnection-Exception is thrown with a status of BluetoothConnection-Exception.UNACCEPTABLE_PARAMS. The error handling code for the L2capClient class checks for a BluetoothConnection-Exception with that status. The BluetoothConnectionException class defines five other constants in addition to UNACCEPTABLE_ PARAMS. These constants describe different reasons that a connection attempt might fail. As shown in the example code, the getStatus() method is used to retrieve the constant that applies to a particular exception.

Once the L2CAPConnection is established, it is passed to the method sendL2capPacketsForEcho(), which sends and receives the bytes over the L2CAP channel. The sendL2capPacketsForEcho() method is inherited from EchoParticipant, which is the next class we examine. This class has two methods. The method sendL2cap-PacketsForEcho() is used by the client for generating the L2CAP packets. The method echoReceivedL2capPackets() is used by the server for echoing back the bytes received from the client.

```
public class EchoParticipant {

  protected StringItem out;
  private int bytesToSend = 100000;

  void sendL2capPacketsForEcho(L2CAPConnection conn)
    throws IOException {
    byte[] sbuf = new byte[conn.getTransmitMTU()];
    byte[] rbuf = new byte[conn.getReceiveMTU()];
    for (int i=0; i < sbuf.length; i++) {
      sbuf[i] = (byte)i;
    }
    int count = 0;
    long start = System.currentTimeMillis();
    while (count < bytesToSend) {
      conn.send(sbuf);
      count += sbuf.length;
      count += conn.receive(rbuf);
      /* Display the bytes sent and received so far */
      out.setText(Integer.toString(count));
    }
    /* Let the echoer know we are done sending bytes */
    conn.send("DONE".getBytes());
    conn.receive(rbuf);
    long end = System.currentTimeMillis();
    out.setText("Done (transferred "+count+" bytes)\n"
      + "Elapsed time " + (end - start)/1000
      + "sec");
  }

  void echoReceivedL2capPackets(L2CAPConnection conn)
    throws IOException {
    byte[] ibuf = new byte[conn.getReceiveMTU()];
    int bytesIn;
    int count = 0;
    for (;;) {
      bytesIn = conn.receive(ibuf);
      byte[] obuf = new byte[bytesIn];
```

```
        System.arraycopy(ibuf, 0, obuf, 0, bytesIn);
        conn.send(obuf);
        if ((bytesIn == 4) &&
          (new String(obuf)).equals("DONE")) {break;}
        count += 2 * bytesIn;
        /* Display the bytes received and sent so far */
        out.setText(Integer.toString(count));
      }
    out.setText("Done (transferred " + count + " bytes)");
  }
}
```

The key parts of both methods are the statements that send and receive L2CAP packets. The method `sendL2capPacketsForEcho()` does

```
    conn.send(sbuf);
```

followed by

```
    count += conn.receive(rbuf);
```

The method `echoReceivedL2capPackets()` reverses the order of these operations.

The method `sendL2capPacketsForEcho()` follows MTU Rule 3, which limits the size of packets sent to a maximum of `getTransmitMTU()`. This is shown in the two statements

```
    byte[] sbuf = new byte[conn.getTransmitMTU()];
```

and

```
    conn.send(sbuf);
```

The method `sendL2capPacketsForEcho()` also follows MTU Rule 4, which recommends allocating a byte array of size `getReceiveMTU()` to receive incoming packets

```
    byte[] rbuf = new byte[conn.getReceiveMTU()];
```

and

```
    count += conn.receive(rbuf);
```

The method `echoReceivedL2capPackets()` also follows MTU Rules 3 and 4, although this is more difficult to see for MTU Rule 3. The relevant statements are

```
bytesIn = conn.receive(ibuf);
byte[] obuf = new byte[bytesIn];
  ...
conn.send(obuf);
```

The size of the byte array sent is based on the size of the byte array received. How do we know that the byte array `obuf` in `conn.send(obuf)` is not larger than `transmitMTU`? Because all MTU values are specified as 672 bytes in both the server and client connection strings, we know that the packet received will not be larger than 672 bytes. We can conclude that `obuf`, the byte array sent, also will be no larger than 672 bytes.

The client knows it is finished sending packets when its count exceeds 100,000 bytes. The `L2capClient` sends a special 4-byte packet corresponding to the ASCII values for the character string "DONE" to inform the `L2capServer` that the client is finished sending bytes.

8.3.2 User Interface for MTU Values

The example code in this section extends the L2CAP echo program of the previous section with a user interface that lets you enter values for `receiveMTU` and `transmitMTU` for both the L2CAP server and the L2CAP client. These MTU values are then used in the connection strings. If an empty value is provided for one of these MTU values, that is, if the field in the user interface is cleared, the corresponding MTU parameter is not included in the connection string passed to `Connector.open()`. This user interface makes it easy to try various combinations of MTUs to see the effect of MTU Rules 1 and 2 and the MTU mismatch pitfall discussed in Section 8.2.3. This user interface also makes it possible to experience the effect of MTU size on the time required for the client and server to transmit 100,000 bytes over L2CAP.

Figure 8.4 shows the user interface for entering MTU values. Both the server and the client use the same user interface. Figure 8.4(A) shows the first screen, which allows the user to enter a value for `receiveMTU`. Figure 8.4(B) shows the second screen, which allows the user to enter a

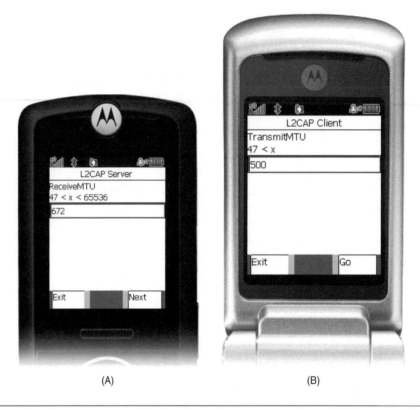

Figure 8.4   User interface for MTU experiments. (A) A `receiveMTU` of 672 bytes is entered for the server. (B) A `transmitMTU` of 500 bytes is entered for the client (emulation only).

value for `transmitMTU`. The values shown in Figure 8.4 are compatible because `transmitMTU` ≤ `receiveMTU` for L2CAP packets sent from the client to the server.

The method `getReceiveMTUFromUser()` shown below creates the display shown in Figure 8.4(A) for entering a value for `receiveMTU`. The constant `L2CAPConnection.MINIMUM_MTU` is used to display the lower bound on legal input values. The `LocalDevice` property `bluetooth.l2cap.receiveMTU.max` is used to display the upper bound on legal input values. The constant `L2CAPConnection.DEFAULT_MTU` is provided as the starting value of the input field.

```
private void getReceiveMTUFromUser(boolean isClient) {
  String maxRecMTUPlus1;
  String maxRecMTU = LocalDevice.getProperty(
    "bluetooth.l2cap.receiveMTU.max");

  if (maxRecMTU == null) {
    maxRecMTUPlus1 = "Unknown";
  } else {
    // Get (max + 1) for display of (min - 1) < x < (max + 1)
    maxRecMTUPlus1 =
      (new Integer(Integer.parseInt(maxRecMTU) +
      1)).toString();
  }

  String initialMTU =
    Integer.toString(L2CAPConnection.DEFAULT_MTU);
  receiveMTUForm = new Form(isClient ? "L2CAP Client" :
    "L2CAP Server");
  String recMtuFieldLabel =
    "ReceiveMTU \n" + (L2CAPConnection.MINIMUM_MTU
    - 1) + " < x < " + maxRecMTUPlus1;
  receiveMTUForm.append(new TextField(recMtuFieldLabel,
    initialMTU, 10, TextField.NUMERIC));
  receiveMTUForm.addCommand(new Command("Exit",
    Command.EXIT, 1));
  receiveMTUForm.addCommand(new Command("Next",
    Command.ITEM, 1));
  receiveMTUForm.setCommandListener(this);
  display.setCurrent(receiveMTUForm);
}
```

We can cause problems for the `L2capServer` if we enter the following MTU values in the user interface:

> Client: `receiveMTU=500`, `transmitMTU=672`
> Server: `receiveMTU=672`, `transmitMTU=500`

These values avoid the MTU mismatch pitfall, so the connection is formed. The client sends 672-byte L2CAP packets to the server. The server can receive these packets because 672 bytes is the same as the server's `receiveMTU`. However, when the server attempts to echo the bytes back to the client, a 672-byte packet is larger than the server's `transmitMTU` of 500 bytes.

There are several options for dealing with this problem. For the sake of simplicity, we adopt an approach that does only one `send(outBuf)` for each `receive(inBuf)` and uses an `outBuf` of size `transmitMTU`. The extra 172 bytes are not echoed back to the client. This allows us to have just one `send()` for every `receive()` and keeps the example code a little simpler. In theory, using an `outBuf` of 672 bytes with all the received bytes should lead to the same result as using an `outBuf` of size `transmitMTU`. The excess 172 bytes should be automatically discarded by the JABWT implementation. However, this would violate MTU Rule 3. Following MTU Rule 3 here makes it clear in the code that the failure to echo all the bytes is intentional.

The shaded statements below show the changes made to the method `echoReceivedL2capPackets()` shown in Section 8.3.1. The shaded statements limit the number of bytes echoed from each incoming L2CAP packet to just `transmitMTU` bytes.

```
void echoReceivedL2capPackets(L2CAPConnection conn)
  throws IOException {
  byte[] ibuf = new byte[conn.getReceiveMTU()];

  int transmitMTU = conn.getTransmitMTU();

  int bytesIn;

  int bytesOut;

  int count = 0;
  for (;;) {
    bytesIn = conn.receive(ibuf);

    bytesOut = Math.min(bytesIn, transmitMTU);
    byte[] obuf = new byte[bytesOut];
    System.arraycopy(ibuf, 0, obuf, 0, bytesOut);
```

```
conn.send(obuf);
if ((bytesIn == 4) &&
    (new String(obuf)).equals "DONE"))  {
    break;
}
```

```
count += bytesIn + bytesOut;
```

```
    /* Display the bytes received and sent so far */
    out.setText(Integer.toString(count));
  }
  out.setText("Done (transferred " + count + " bytes)");
}
```

8.3.3 L2CAP Clients and Servers Have the Same Capabilities

The example code in this chapter might leave the erroneous impression that L2CAP server applications have to be passive and are incapable of initiating communications. This is not the case. Although the client initiates the L2CAP connection, once that connection is formed, both sides have access to an instance of an L2CAPConnection, so both sides have the same capabilities. Clients and servers can both send packets whenever they want to. An easy experiment that illustrates this point is to exchange these two statements in the example code:

```
sendL2capPacketsForEcho(conn);
```

and

```
echoReceivedL2capPackets(conn);
```

This exchange changes the example code so that instead of the server echoing the packets sent by the client, the client echoes the packets sent by the server.

8.3.4 Flow Control

Initial versions of the Bluetooth specification included no L2CAP flow control. Bluetooth specification version 1.2 introduced L2CAP flow control, but made it optional. So Bluetooth protocols and profiles that use L2CAP typically provide their own flow control. Let's consider the echo example of this chapter from the point of view of flow control. Suppose the client device is capable of sending L2CAP packets much faster than the server device is capable of echoing them back. Because our example code waits for a packet to be returned before it tries to send the next packet, the client is paced by the server's ability to echo the packets. This should keep the client from getting ahead of the server and overflowing the server's buffers.

However, suppose that instead of a symmetric, two-way echoing application, the data transfer is one way. In that case, a different flow control scheme is required. In addition to the L2CAP packets that transmit data and travel in one direction it would be necessary to send back L2CAP packets containing control signals to stop and start the data flow.

The next code example illustrates a one-way data transfer using a credit-based flow control scheme.

The code for the complete example is too lengthy to be shown here, so only selected methods involved with credit-based flow control are shown. The complete code is available at the Web site for this book www.mkp.com. Credit-based flow control is used in the Bluetooth specification for RFCOMM and the Hardcopy Cable Replacement Profile. In the credit-based flow control scheme illustrated in this example, the L2CAP server starts by issuing four credits to the L2CAP client. The client can send as many L2CAP packets as it has credits, so the client can then send four packets to the L2CAP server. When the client's credit count reaches zero, the client must stop sending L2CAP packets and wait for additional credits from the server.

In the `CreditBased1WayXfer` class shown below, the key section that accomplishes the flow control is

```
if (availableCredits > 0) {
    conn.send(sbuf);
    availableCredits--;
```

The L2CAP client can use the JABWT `send()` method to send an L2CAP packet to the server as long as it has available credits. However, each packet sent uses up a credit. When the credits reach zero, the L2CAP client has to stop sending data until more credits are received.

```
public class CreditBased1WayXfer {
  // Number of L2CAP packets the receiver has
  // authorized to be sent
  int availableCredits;

  protected StringItem out;
  private int bytesToSend = 50000;

  void sendL2capPackets(L2CAPConnection conn) throws
    IOException {

    boolean sentDone = false;
    byte[] sbuf = new byte[conn.getTransmitMTU()];
    int receiveMTU = conn.getReceiveMTU();
    byte[] rbuf = new byte[receiveMTU];

    for (int i=0; i < sbuf.length; i++) {
      sbuf[i] = (byte)i;
    }

    int count = 0;
    long start = System.currentTimeMillis();

    // listen for credits authorizing sending packets
    receiveCredits(conn, receiveMTU);
    while (count < bytesToSend) {
      if (availableCredits > 0) {
        conn.send(sbuf);
        availableCredits--;
        count += sbuf.length;

        // Display the number of bytes sent so far
        out.setText(Integer.toString(count));
      }

      maybeReceiveCredits(conn, receiveMTU);
    }

    // Let the receiver know we are done sending bytes
    while (!sentDone) {
      if (availableCredits > 0) {
        conn.send("DONE".getBytes());
        sentDone = true;
      } else {
```

```
      maybeReceiveCredits(conn, receiveMTU);
    }
  }
  long end = System.currentTimeMillis();
  out.setText("Done (transferred "+count+" bytes)\n"
    + "Elapsed time " + (end - start)/1000 + "sec");
  }
}
```

The sendL2capPackets() method above uses the two application methods receiveCredits() and maybeReceiveCredits() to listen for L2CAP packets that deliver additional credits from the server. The definition of the receiveCredits() method is shown next. This method reads an L2CAP packet from the server and interprets the byte array in that packet as an integer. That integer is added to available-Credits to increase the credits available for use by the client.

```
// Read an L2CAP packet. If it has four bytes, then
// interpret those bytes as new credits for sending
// L2CAP packets.
void receiveCredits(L2CAPConnection conn,
  int receiveMTU) {
  int incomingBytes;
  int newCredits;
  byte[] rbuf = new byte[receiveMTU];
  try {
    incomingBytes = conn.receive(rbuf);
  } catch (IOException ignore) {
    return;
  }
  // assume four bytes are used to encode new
  // credits
  if (incomingBytes != 4) {
    return;
  }
  availableCredits +=
    CreditBased1WayXfer.byteArray2Int(rbuf);
}
```

```
// Convert a four-byte array to an int. The byte
// array is assumed to have a big Endian
// byte order.
public static int byteArray2Int(byte[] argBytes) {
  int result = 0;
  // big-endian conversion
  for (int i = 0, j = 0; i < 4; i++, j++) {
    result = result + (((int)argBytes[i] << 24)
      >>> (j * 8));
  }

  return result;
}
```

The L2CAP server issues an additional credit to the client only when the server frees up buffer space to hold one additional L2CAP data packet. It is unclear how long it takes the server to process one of the previously sent L2CAP packets, free up the space needed to receive another packet from the client, and send an L2CAP packet back to the client to issue the additional credit. Because receiveCredits() uses the blocking JABWT method receive() to read an L2CAP packet from the server, there is always the risk that a call to receive() can become stuck waiting for this L2CAP packet from the server. The credit-based flow control scheme is meant to block only when the client has no more credits, so we want the client to listen for additional credits without blocking.

The key to keeping the client from getting stuck in receive() is to use the JABWT ready() method to test whether an L2CAP packet is available for the client to read. If ready() returns true, receive() returns an L2CAP packet without blocking. The client's maybeReceive-Credits() method is shown next. It uses the ready() method to check for additional credits issued by the server without blocking.

```
// If there is an L2CAP packet waiting to be read,
// then call receiveCredits. Otherwise return
// without blocking.
void maybeReceiveCredits(L2CAPConnection conn,
  int receiveMTU) {
```

```
try {
  if (conn.ready()) {
    receiveCredits(conn, receiveMTU);
  }
} catch (IOException ignore) {
  }
}
```

All of the example code we have looked at so far for credit-based flow control has been client code. The L2CAP server in this example has two threads: one thread to move the L2CAP packets received from the client to a buffer storage location and a second thread to process the buffered L2CAP packets and issue credits.

The receiveL2capPackets() method shown next is used in the thread that receives the incoming L2CAP packets and stores them. It is derived from the echoReceivedL2capPackets() used in the echo examples earlier in this chapter. The receiveL2capPackets() method has been modified to store the bytes obtained from an L2CAP packet sent by the client to one of four L2capPacketBuffers maintained by the server. An L2capPacketBuffer has room to store getReceiveMTU() bytes. These bytes are processed later by the second thread. The code in echoReceivedL2capPackets() for counting incoming bytes and for echoing those bytes back to the client has been removed from receive-L2capPackets().

```
public class CreditBased1WayXfer {

...

  void receiveL2capPackets(L2CAPConnection conn,
    L2capBuffers buffers,
    CreditIssuer issuer)
    throws IOException {
    byte[] ibuf = new byte[conn.getReceiveMTU()];
    L2capPacketBuffer packetBuffer;
    int transmitMTU = conn.getTransmitMTU();
    int bytesIn;
    for (;;) {
```

```
    packetBuffer =
      buffers.nextAvailablePacketBuffer();
  if (packetBuffer != null) {
    bytesIn = conn.receive(ibuf);
    packetBuffer.storeBytes(ibuf, bytesIn);
    if ((bytesIn == 4) &&
      (new String(ibuf, 0,
      bytesIn)).equals("DONE")) {
      break;
    }
  } else {
    if (conn.ready()) {
      System.out.println("Should not get here. No " +
        "L2capPacketBuffer available " +
        "to receive incoming packet.");
    }
  }
}
issuer.setDoneProcessing();
out.setText("Done (transferred " + issuer.count +
  " bytes)");
  }
...
}
```

The following example code shows the `run()` method for the `Credit-Issuer` thread that processes the buffered packets and issues credits to the client. The first thing that happens in the `CreditIssuer.run()` method is that an L2CAP packet is sent to the client containing a number represented as a byte array. The number in the L2CAP packet is the number of starting credits that the server issues to the client. This number is determined by the number of client packets that the server has buffer space available to receive.

The `CreditIssuer` then continuously loops over the buffers to process any packets that have been received and stored there by the other server thread. The only processing of packets done in the example is to count the number of bytes in each packet and keep a total count of bytes received from the client over this L2CAP channel.

```java
public class CreditIssuer implements Runnable {
...
  public void run() {
    L2capPacketBuffer packetBuffer;
    int freedBuffers;
    // Issue one credit for each L2capPacketBuffer
    int totalCredits =
      L2capBuffers.NUMBER_OF_PACKET_BUFFERS;
    byte[] issueCreditsPacketPayload =
      CreditBased1WayXfer.int2ByteArray(totalCredits);
    try {
      conn.send(issueCreditsPacketPayload);
    } catch (IOException e) {
      System.out.println("IOException when issuing "
        + "initial credits");
      return;
    }

    while (!doneProcessing) {
      freedBuffers = 0;
      while ((packetBuffer =
        buffers.nextUsedPacketBuffer())!= null) {
        count += packetBuffer.getNumBytesStored();
        packetBuffer.eraseStoredBytes();
        freedBuffers++;
      }
      if (freedBuffers > 0) {
        try {
          conn.send(CreditBased1WayXfer.int2ByteArray(
            freedBuffers));
        } catch (IOException e) {
          System.out.println("IOException " +
            e.getMessage());
        }
      }
```

```
      try {
        Thread.sleep(sleepTime);
      } catch (InterruptedException ignore) {
      }
    }
  }...
}
```

After the bytes in a buffered L2CAP packet have been counted, the `CreditIssuer` thread calls the application method `eraseStored-Bytes()` to make this `L2capPacketBuffer` available to store future L2CAP packets. `CreditIssuer` keeps track of how many buffers it has freed up and uses the following statement to send an L2CAP packet to the client to issue one additional credit for each buffer freed:

```
conn.send(CreditBased1WayXfer. int2ByteArray(freedBuffers));
```

The statement `Thread.sleep(sleepTime)` at the bottom of the `while` loop makes it possible to introduce an arbitrary delay into the `Credit-Issuer` thread. Experimenting with various delays shows how credit-based flow control adjusts the client's data-transmission rate to match the server's packet-processing rate.

8.4 Summary

L2CAP is one of three APIs for Bluetooth communication that are available to JABWT applications. JABWT provides a packet-based API for L2CAP as opposed to the stream-based APIs available for serial port and OBEX.

L2CAP communications are the right choice for an application if

- The application implements a Bluetooth profile that uses the L2CAP protocol and that Bluetooth profile does not use one of the higher-level protocols RFCOMM or OBEX, or
- The application implements a new custom protocol that is packet oriented.

The `L2CAPConnection` interface provides methods for sending and receiving L2CAP packets over an L2CAP channel. Applications can use the connection string parameters `receiveMTU` and `transmitMTU` to define their requirements for maximum payload sizes of the L2CAP packets. This chapter presents four rules regarding MTU values for use in JABWT programs.

L2CAP flow control is an optional part of the Bluetooth specification, so JABWT applications that use the L2CAP API should provide their own flow control. Without some form of flow control, L2CAP applications could encounter packet loss or deadlock. This chapter provides example code for two flow control schemes:

- Simple flow control that waits for a response to packet n before sending packet $n + 1$
- Credit-based flow control.

9 Push Registry

This chapter covers how to utilize the Push Registry with JABWT. In particular, this chapter covers the following:

- A brief introduction to the Push Registry
- How to statically/dynamically register and unregister connection strings
- How to update service records on a Push Registry service

9.1 Overview

Most Java ME devices support only running a single Java MIDlet at a time. (This may be changing in the next couple of years as a result of devices getting more capabilities, memory, and resources.) Because of the resource-constrained nature of a Java ME device, running multiple background MIDlets is typically not supported. Lack of support for running MIDlets in the background made it difficult to support applications that would simply wait to receive a connection or message from a remote device. This issue came to a head around the time JSR-120 (Wireless Messaging) was being completed. Developers were looking for ways to pass messages to Java ME devices that may be consumed by MIDlets. Historically, MIDP 1.0 devices assumed a pull method of getting data, requiring the MIDlet to be running to receive data.

When the original JABWT specification was completed, this issue became even more pronounced. While the JABWT specification did not solve the problem, it did expect to see such a capability defined in the future. In JABWT, the ability to launch a MIDlet at any time based on an input connection was referred to as a connect-anytime service. In the 1.1

version of the JABWT specification, connect-anytime services were replaced with the Push Registry.

The Push Registry was introduced in the MIDP 2.0 specification. The Push Registry provided a way for MIDlets to register to accept connections even when the MIDlet is not running. The Push Registry resides in the Java Application Manager (JAM) (see Figure 9.1). The JAM is responsible for installing, uninstalling, starting, pausing, and stopping Java MIDlets on a Java ME device. When a MIDlet registers a Push Registry request with the JAM, the MIDlet is stating that it wishes to start when a connection is made to the device with the characteristics provided in the request.

When the Push Registry receives a connection from another device, the Push Registry notifies the JAM. The JAM starts the MIDlet that registered for the specific request. After the MIDlet starts, the MIDlet can then read and write data over the connection with the other device. After processing the request, the MIDlet ends and waits for the JAM to start the MIDlet on another connection request.

Prior to the Push Registry, MIDlet developers had to rely on users to start a MIDlet to make the connections that the developer wanted to make. For example, a MIDlet developer that wished to synchronize with a desktop would have to have the MIDlet establish the connection. With the introduction of the Push Registry, a MIDlet developer is able to

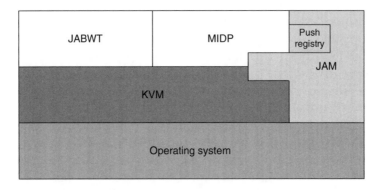

Figure 9.1   The JAM and Push Registry in a MIDP 2 device.

register a synchronization service. Now, the desktop can discover and connect to the Bluetooth device. The synchronization can then start without the user having to start the MIDlet, thus allowing devices to complete tasks more seamlessly.

9.2 API Capabilities

JABWT does not define any new APIs to support the Push Registry, but rather utilizes the APIs defined in the MIDP 2.X specification. As a result, the Push Registry only exists on devices that support MIDP 2.X or greater specifications and claim support for Bluetooth connections in the Push Registry. JABWT defines a standard way to specify requests to the Push Registry. JABWT also specifies how the Push Registry creates service records. JABWT does not specify how the Push Registry works with non-Bluetooth protocols such as OBEX over TCP or OBEX over IrDA.

With the Push Registry, JABWT allows MIDlets to statically and dynamically register connect-anytime services. Static registration occurs when the MIDlet suite is installed. To enable static registration, the JAD file of the MIDlet is modified to specify the connection string of the service and provides a list of devices that may connect to the service. Dynamic registration occurs after the MIDlet is installed. To register a service dynamically, the MIDlet must be started by the user. While the MIDlet is running, the MIDlet interacts with the Push Registry with a set of APIs to specify the connection string and list of devices that may connect to the service. A MIDlet may register a connection string that starts any MIDlet in the MIDlet suite.

During registration of the service, a Bluetooth service record is created for the service. The service record is created in the same way specified in Chapter 7 based on the connection string provided to the Push Registry. After creating the service record, the service record is added to the SDDB. For a statically registered service, the service record is activated and the Push Registry begins listening for connection requests (see Figure 9.2).

Service records are handled slightly different for dynamically registered connection strings. Dynamically registered connection strings do

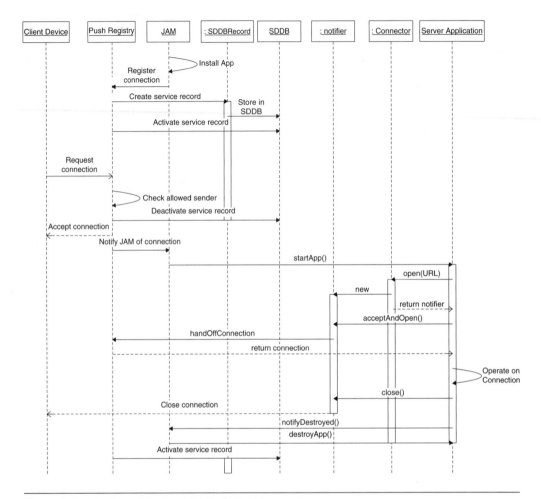

Figure 9.2   Static registration of a Push Registry service.

not have their service records activated until after the MIDlet that registered the connection string ends. Once the MIDlet ends, the Push Registry begins listening for connections for the service (see Figure 9.3). The major difference between static and dynamic registration is shown in the gray box in Figure 9.3.

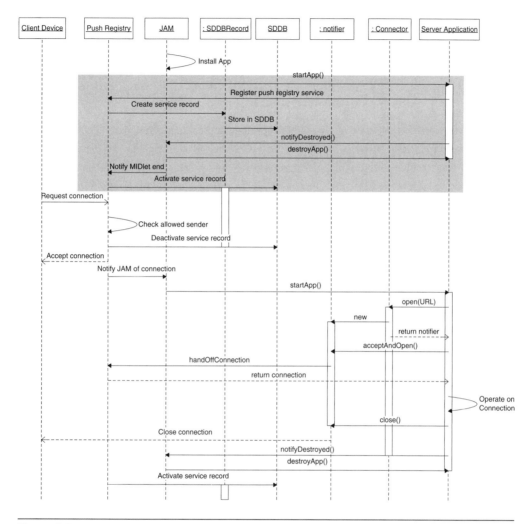

Figure 9.3   Dynamic registration of a Push Registry service.

The service record created by the Push Registry may also be modified. While modifications are not allowed to affect the connection string, new data elements may be added to provide additional information to devices that may be looking for the service.

9.3   Programming with the API

MIDP 2.0 defines two ways to interact with the Push Registry: installation time (i.e., statically) or run time (dynamically). Regardless of which approach is used, three pieces of information are required to register a MIDlet in the Push Registry.

1. The server connection string that describes the service being provided.

2. The string name of the MIDlet that will handle connections to the service. The MIDlet must be in the MIDlet suite that is registering the connection.

3. The list of devices that are allowed to use the service and/or the list of devices that are not allowed to use the service. These two lists combined are known as the `AllowSender` parameter.

The `AllowSender` parameter is a string that describes requirements that must be met to allow a device to connect to the service. In the `Allow-Sender` string, the string may specify Bluetooth device addresses that may connect (with wild cards), minimum security requirements that must be supported, and a list of Bluetooth devices that may not use the service. To support these three levels of control, the string is broken into two parts. The first part specifies a list of devices that may connect to the service by specifying their Bluetooth address in a semicolon-separated list. Within this list, the "*" and "?" wild cards may be used. Table 9.1 provides some examples of `AllowedSender` strings that specify devices that may connect to the service.

To support higher levels of security, security requirements may also be specified in the `AllowSender` string. To specify that a device must be

Table 9.1   Examples of Valid `AllowSender` Strings

AllowSender Value	Meaning
00E000482312	Only allow the Bluetooth device with address of 00E000482312 to connect
00E000??????;001122334455	Allow devices that begin with 00E000 to connect and the device with address 001122334455
*	Allow all devices to connect

authenticated prior to using the service, the ";authenticated" string is added after the Bluetooth address. Likewise, ";authorized" is added after the Bluetooth address when the device must be authorized to use a service. Both ";authenticated" and ";authorized" strings do not need to be listed for a Bluetooth address, as authorization implies authentication. To specify that all devices must be authorized to use the service, the following string would be used:

```
*;authorized
```

To specify that device 00E000482312 must be authenticated, the following string would be used:

```
00E000482312;authenticated
```

Specifying authenticated or authorized in AllowSender strings does not cause the operations to be performed, but rather determines if the condition is true. In other words, specifying ";authenticated" in the AllowSender string is not equivalent to "authenticate=true" in the connection string. Specifying "authenticate=true" in the connection string allows an incoming connection to pass the ";authenticated" test in the AllowSender string. If the connection has not been authenticated, putting ";authenticated" in the AllowSender string does not tell the system to start the authentication process.

Note that the terms used in AllowSender strings are different from the terms used in connection strings. AllowSender strings use authenticated and authorized, whereas connection strings use authenticate and authorize.

The second part of the AllowedSender string specifies devices that are not allowed to use the service. These devices are blacklisted from the service by adding the "blacklist=" string and the device addresses. The "*" and "?" wild cards may be used in a blacklist address. For example, the following string allows all devices except for devices that start with 001122:

```
*;blacklist=001122*
```

By combining the allowed list with the blacklist, a developer is able to restrict access to services based on security requirements and the Bluetooth address of a device. Table 9.2 shows some examples of the AllowSender parameter using the blacklist option.

Table 9.2 Examples of `AllowSender` Parameter with a `blacklist` Value

AllowSender Parameter	Meaning
;authenticated;blacklist=00E012	Devices with a Bluetooth address starting with 00E012 are not allowed to use the service. All other devices are allowed as long as the devices are authenticated.
0011223344??;blacklist=001122334456	Devices whose Bluetooth address starts with 0011223344 may use the service except for the device whose address is 001122334456.
00E0*;blacklist=00E00023030?	Devices that start with 00E0 may use the service except for devices that start with 00E00023030.

9.3.1 Static Registration

To register a Push Registry connection at installation time, a new attribute is added to the JAD file for the MIDlet suite. The `MIDLET-PUSH-<n>` attribute, where `<n>` is a positive integer, describes each Push Registry connection string to add. (The attribute is not case sensitive.) Within the JAD, multiple services may be specified by incrementing `<n>`. The first-ribute must be `MIDLET-PUSH-1`. The attribute requires the three parameters mentioned previously separated by commas. For example, to specify an RFCOMM service that will be fulfilled by a `TestMIDlet` class that accepts a connection from all devices, the following string may be used:

```
MIDlet-Push-1: btspp://localhost:12412421, TestMIDlet, *
```

In Chapter 4, the `EchoServer` MIDlet was created. As part of the process of compiling and testing the MIDlet, the Sun Wireless Toolkit created a JAD file that was used by the Motorola LaunchPad to run the MIDlet. The JAD file from Chapter 4 has been modified to support the Push Registry. The modified JAD file (modification shown in gray and must appear on a single line unlike shown here) registers a push connection at installation for a connect-anytime service that uses RFCOMM with a service name of "Echo Server." The service allows all devices to connect to it.

```
MIDlet-1: EchoServerPush, EchoServerPush.png,
   com.jabwt.book.EchoServer
MIDlet-Jar-Size: 2796
MIDlet-Jar-URL: EchoServerPush.jar
```

```
MIDlet-Name: EchoServerPush
```

```
MIDlet-Push-1: btspp://localhost:123456789ABCDE;
  name=Echo Server, com.jabwt.book.EchoServer, *
```

```
MIDlet-Vendor: JABWT Book
MIDlet-Version: 1.0
MicroEdition-Configuration: CLDC-1.1
MicroEdition-Profile: MIDP-2.0
```

A single MIDlet may handle multiple Push Registry connection strings. Connection strings are registered when the MIDlet suite is installed and remain registered until the MIDlet suite is uninstalled. The installation will fail if the Push Registry cannot meet the request in the JAD file. A request may not be fulfilled because the connection string is not valid, the device does not support JABWT push, the MIDlet is not in the MIDlet suite, or the `AllowSender` parameter is malformed.

9.3.2 Dynamic Registration

Dynamic registration allows a MIDlet to register and unregister a Push Registry service while a MIDlet is running. To register while a MIDlet is running, the `javax.microedition.io.PushRegistry` class is used. The `registerConnection()` function is used to register the connection. The connection string, MIDlet class name and `AllowSender` string are passed to the `registerConnection()` function to dynamically register a service. Depending on the implementation, the Push Registry might not start accepting connections for the service until after the MIDlet terminates. Table 9.3 shows exceptions that may be thrown by a call to `registerConnection()`.

The `PushRegistry` class also provides a way to retrieve all the connection strings that have registered with the Push Registry for the current MIDlet suite. The `listConnections()` method provides a `String` array listing every connection string that has registered. The `listConnections()` method has one argument to specify if only those connection strings that are active in the Push Registry or all connection strings should be returned.

Table 9.3 Exceptions That May Be Thrown by `registerConnection()`

Exception	Reason
`IllegalArgumentException`	If the connection or `AllowSender` string is not valid
`ConnectionNotFoundException`	If the device does not support push delivery for JABWT
`IOException`	If insufficient resources are available to handle the registration request
`ClassNotFoundException`	If the MIDlet class name is not in the MIDlet suite

To show how JABWT and the Push Registry work together, a new MIDlet will be created to add push support to the `EchoServer` MIDlet suite that was created in Chapter 4. The `RegisterEchoServer` MIDlet will first determine if a connection string has already been registered with the Push Registry for the MIDlet suite. If a connection string has not already been registered, the `RegisterEchoServer` MIDlet will attempt to register a connection string with the Push Registry. The user will be notified if the connection was registered successfully or the reason for the failure. After the notification is displayed, the `RegisterEcho-Server` MIDlet will close.

```
package com.jabwt.book;
import java.lang.*;
import java.io.*;
import javax.microedition.midlet.*;
import javax.microedition.lcdui.*;
import javax.microedition.io.*;
import javax.bluetooth.*;
public class RegisterEchoServer extends BluetoothMIDlet {
  private static final String CONN_STRING =
    "btspp://localhost:123456789ABCDE;" +
    "name=Echo Server"
  /**
   * Register a connection with the Push Registry as
   * long as it has not already registered.
   */
```

```
public void run() {
  Alert msg = null;
  String[] connStrings =
    PushRegistry.listConnections(false);

  if ((connStrings == null) ||
    (connStrings.length == 0)) {
    msg = registerEchoServer();
  } else {
    msg = new Alert("Error",
      "The connection string is " +
      "already registered.", null,
      AlertType.ERROR);
  }

  msg.setCommandListener(this);
  Display.getDisplay(this).setCurrent(msg);
}

/**
 * Registers a connection with the Push Registry.
 *
 * @return an alert with the status of the request
 */
private Alert registerEchoServer() {
  Alert msg = null;
  try {
    PushRegistry.registerConnection(CONN_STRING,
      "com.jabwt.book.EchoServer", "*");
    msg = new Alert("Register",
      "Service successfully registered",
      null, AlertType.CONFIRMATION);
  } catch (ConnectionNotFoundException e) {
    msg = new Alert("Not Supported",
      "Bluetooth Push Registry not supported",
      null, AlertType.ERROR);
  } catch (IOException e) {
    msg = new Alert("Failed",
      "Failed to register connection " +
```

```
        "(IOException: " + e.getMessage() +
        ")", null, AlertType.ERROR);
    } catch (ClassNotFoundException e) {
      msg = new Alert("Failed",
        "Failed to register service",
        null, AlertType.ERROR);
    }

    return msg;
  }
}
```

The Motorola SDK does not support JABWT connections in its Push Registry implementation. As a result, the Sun Wireless Toolkit is used to test the code. Because the `EchoServer` and `RegisterEchoServer` MIDlets are in the same MIDlet suite, the user must select to launch the `RegisterEchoServer` MIDlet (see Figure 9.4A). Most Push Registry implementations will require the user to accept the registration request as shown in Figure 9.4B. Upon successful registration, the `RegisterEchoServer` MIDlet displays a successful registration request (see Figure 9.4C). Using the `listConnections()` method, the `RegisterEchoServer` is able to verify whether a connection string has already been registered. If a connection string has already been registered, an error message is displayed (see Figure 9.4D).

The `PushRegistry` class also provides a way to unregister a connection using the `unregisterConnection()` method. The argument of the `unregisterConnection()` method is the connection string used to register the connection. The connection string must be exactly the same as the string used in the `registerConnection()` method. The method returns `true` if the connection string was removed from the Push Registry. The method returns `false` if the connection string was not registered in the Push Registry.

To show the use of the `unregisterConnection()` method in example code, the `RegisterEchoServer` MIDlet will be modified to unregister the connection if the connection string was in the Push Registry.

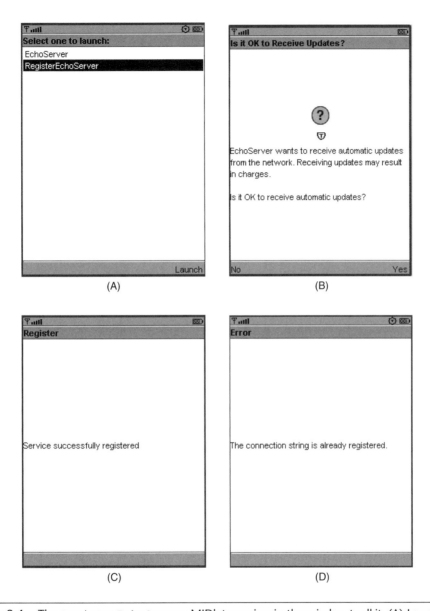

Figure 9.4 The `RegisterEchoServer` MIDlet running in the wireless toolkit. (A) Launching the `RegisterEchoServer` MIDlet. (B) Confirmation of registration. (C) Successful registration. (D) Failure after the connection string has already been registered.

```
public class RegisterEchoServer extends BluetoothMIDlet {
  ...
  public void run() {
    Alert msg = null;
    String[] connStrings =
      PushRegistry.listConnections(false);
    if ((connStrings == null) ||
      (connStrings.length == 0)) {
      msg = registerEchoServer();
    } else {
```

```
      msg = unregisterEchoServer();
```

```
    }
    msg.setCommandListener(this);
    Display.getDisplay(this).setCurrent(msg);
  }
```

```
  /**
   * Unregisters the connection with the Push
   * Registry.
   *
   * @return an alert with the status of the request
   */
  private Alert unregisterEchoServer() {
    Alert msg = null;
    if (PushRegistry.unregisterConnection(
      CONN_STRING)) {
      msg = new Alert("Success",
        "Successfully Unregistered Connection",
        null, AlertType.CONFIRMATION);
    } else {
      msg = new Alert("Failed",
        "Unregister failed", null,
```

```
        AlertType.ERROR);
   }

   return msg;
 }
```

   . . .

 }

9.3.3 Retrieving the Connection

When another device attempts to connect to a service registered via the Push Registry, the Push Registry will accept the connection and ask the JAM to start the MIDlet specified in the registration. If the remote device sends data prior to the MIDlet starting, the JAM is responsible for buffering data until the MIDlet is able to retrieve it. After starting the registered MIDlet, the MIDlet is able to retrieve the connection started by the Push Registry by calling `Connector.open()` with the *exact same string* as the string provided in the Push Registry registration. The `Connection` object returned by `Connector.open()` can then be used to communicate to the remote device over the appropriate protocol.

   To continue with the example from the previous section, only minimal changes need to be made to the `EchoServer` MIDlet in order to process connections from the Push Registry. The following code shows the `EchoServer` MIDlet from Chapter 4 modified to remove the `for` loop, close the `notifier` object when done, and destroy the MIDlet after processing the request. (Note: Neither the Sun Wireless Toolkit nor the Motorola SDK is currently able to start a MIDlet from a JABWT Push Registry connection.)

```
public class EchoServer extends BluetoothMIDlet {

   . . .

   public void run() {
      // Create the output Form and set it to be the
      // current Displayable
      Form msgForm = new Form("Echo Server");
```

```
msgForm.addCommand(new Command("Exit",
  Command.EXIT, 1));
msgForm.setCommandListener(this);
Display.getDisplay(this).setCurrent(msgForm);
try {
  //Create the notifier object
  StreamConnectionNotifier notifier =
    (StreamConnectionNotifier)
    Connector.open("btspp://localhost:123456789ABCDE;"
    + "name=Echo Server");

  // Display the connection string on the Form
  displayConnectionString(msgForm, notifier);

  // Continue accepting connections until the MIDlet
  // is destroyed

  // for (;;) {

  StreamConnection conn = notifier.acceptAndOpen();
  OutputStream output = conn.openOutputStream();
  InputStream input = conn.openInputStream();

  // Continue reading the input stream until the
  // stream is closed. Display the data on the
  // screen and write it to the output stream.
  byte[] data = new byte[10];
  int length = 0;
  while ((length = input.read(data)) != -1) {
    msgForm.append(new String(data, 0, length));
    output.write(data, 0, length);
    output.flush();
  }
  // Close the streams and the connection
  output.close();
  input.close();
  conn.close();
```

```
        //}
        notifier.close();
        notifyDestroyed();
```

```
    } catch (IOException e) {
      msgForm.append("IOException: " + e.getMessage());
    }
  }
}
}
```

After the Push Registry accepts a connection, the Push Registry will stop processing requests to connect to the service until the MIDlet started has closed the notifier object and exited. If the service must be disabled temporarily (i.e., stop processing Push Registry requests), the service may close the server connection object returned by `Connector.open()` and not close the MIDlet.

9.3.4 Lifetime of a Service Record

Chapter 7 covers the lifetime of a service record in a traditional application. When the Push Registry is used, the lifetime of the service record is slightly different. Figures 9.2 and 9.3 hinted at some of the differences, including the fact that a service record is deactivated while a MIDlet is processing a Push Registry request. While there are differences in the lifetime of a service record, the majority of the facts about service records remain the same and the differences are a matter of common sense.

A service record is created and placed in the SDDB once the Push Registry accepts the registration. For static registration, this occurs during the installation process. If the Push Registry is unable to create the service record in the SDDB, installation of the MIDlet suite fails. When a MIDlet registers a connection dynamically, the service record is created and placed in the SDDB before the `PushRegistry.register-Connection()` method returns. If the `registerConnection()` function fails to place the service record in the SDDB, the method throws a `ServiceRegistrationException`. The service record remains in the SDDB until the MIDlet suite is uninstalled or until the MIDlet suite unregisters the service with the Push Registry using `PushRegistry.unregisterConnection()`.

Once the service record is placed in the SDDB, a MIDlet may retrieve the service record in the same way as specified in Chapter 7 using the `LocalDevice.getRecord()` function. Once the service record is retrieved, it can be modified and then updated using the `LocalDevice.updateRecord()` method.

To show how to modify the record, the `RegisterEchoServer` MIDlet will be modified to modify the service record to add a service description to the default service record created when the "Echo Server" service was registered. To update the record, the `registerEchoServer()` method is modified to retrieve the `StreamConnectionNotifier` object in order to get the service record for the service. Once the service record is retrieved, the service record is modified to add the ServiceDescription attribute. Finally, the record is updated in the SDDB by calling `updateRecord()`.

```
public class RegisterEchoServer extends BluetoothMIDlet {

  ...

  private Alert registerEchoServer() {
    Alert msg = null;
    try {
      PushRegistry.registerConnection(CONN_STRING,
      "com.jabwt.book.EchoServer", "*");
```

```
      // Retrieve the notifier object associated
      // with the connection string
      StreamConnectionNotifier notifier =
        (StreamConnectionNotifier)Connector.open(
        CONN_STRING);

      LocalDevice local = LocalDevice.getLocalDevice();
      // Get the service record from the local device
      ServiceRecord record = local.getRecord(notifier);

      // Add ServiceDescription attribute
      record.setAttributeValue(0x0101,
```

```
          new DataElement(DataElement.STRING,
          "This app echoes back messages sent to it."));
     // Update the service record in the SDDB
     local.updateRecord(record);
     notifier.close();
```

```
        msg = new Alert("Register",
          "Service successfully registered",
          null, AlertType.CONFIRMATION);
      } catch (ConnectionNotFoundException e) {
        msg = new Alert("Not Supported",
          "Bluetooth Push Registry not supported",
          null, AlertType.ERROR);
      } catch (IOException e) {
        msg = new Alert("Failed",
          "Failed to register connection " +
          "(IOException: " + e.getMessage() +
          ")", null, AlertType.ERROR);
      } catch (ClassNotFoundException e) {
        msg = new Alert("Failed",
          "Failed to register service",
          null, AlertType.ERROR);
      }
      return msg;
    }
  }
```

9.4 Conclusion

The Push Registry provides support for the connect-anytime service
mentioned in JABWT. The Push Registry was added as part of the MIDP
2 specification to handle incoming connection requests without the
need to keep a MIDlet running all the time. The Push Registry supports
static and dynamic registration. Static registration is done by adding an

entry to the JAD file. To register a connection string dynamically, a MIDlet uses the `PushRegistry` API.

While there are some subtle differences, most of the information learned in previous chapters still applies to Push Registry. When registering a connection with the Push Registry, the same connection strings are used with `Connector.open()` that were described in Chapter 4 (RFCOMM), Chapter 5 (OBEX), and Chapter 8 (L2CAP). The Push Registry also allows a developer to restrict devices that may use the service via the `AllowSender` parameter that is passed in with the registry request.

Once the connection string is registered with the Push Registry, a service record is created based on the connection string. The Push Registry places the service record in the SDDB and activates the service record. The service record may be modified by a MIDlet to provide more information to clients that may wish to use the service. When a client connects to the service, a device may deactivate the service record and stop accepting connections for the service.

10 CHAPTER Closing Remarks

Software standards often are vital to the success of communications technologies. An effective software standard will encourage development of a number of successful applications. Java Specification Request-82, developed by the Java Community Process, standardized the Java APIs for Bluetooth Wireless Technology. JABWT makes it possible to write an application once and then run the application on any Java-enabled device that supports JABWT. Because JABWT was developed with the participation of several companies that develop Bluetooth stacks, we believe it will be possible to implement JABWT in conjunction with a wide variety of Bluetooth stacks. This phenomenon represents a significant change in the way Bluetooth applications will be written and fielded. Because there has been no standard API for Bluetooth stacks, each stack has defined its own API for use by Bluetooth applications. As a consequence, Bluetooth applications have been written to run on a particular Bluetooth stack, and considerable effort has been required to convert that application to run on another Bluetooth stack.

JABWT does not change the fact that Bluetooth stacks all have their own proprietary APIs. JABWT encourages application developers to write their applications to standard JABWT rather than writing them for a particular Bluetooth stack. As device manufacturers adopt JABWT implementations for their Bluetooth devices, JABWT applications will be able to run on those JABWT devices with little or no porting effort on the part of application developers. The different APIs used by the Bluetooth stacks on these devices will be hidden behind the common, standardized API provided by JABWT. The current proliferation of Java ME devices has demonstrated the effectiveness of this strategy and the benefits for Java ME developers. JABWT make it possible for Bluetooth application developers to experience these same benefits. There are currently a sizable number of commerically available devices which support JABWT. The devices include selected mobile phones from Motorola, Nokia, Sony Ericsson, Samsung, and others.

One of the goals of JABWT is to allow third-party vendors to write Bluetooth profiles in the Java language on top of JABWT. Companies have already created Bluetooth profiles using JABWT, especially over OBEX.

JABWT was defined with the participation of many individuals from many different companies. Participation of individuals with different backgrounds helped create a robust specification. The members' expertise runs the entire gamut of topics—Bluetooth hardware, Bluetooth protocol stack, Java ME implementation, Java programming language, OBEX, middleware, and mobile devices design. The JSR-82 effort was a true collaboration and unification of two different industries.

The work completed under JSR-197 allows Java SE devices, such as laptops, to incorporate JABWT. Java SE implementations of JABWT will make the API available to a much larger set of users. It makes logical sense to make it possible for Java SE devices to incorporate JABWT, as Java SE devices are all potential Bluetooth devices.

As we move forward, some newer protocols such as BNEP and profiles such as PAN, which could be widely used in Bluetooth devices, could prompt extending JABWT. Voice- and telephony-related topics were not considered in the current version of JABWT, but they may be considered in the next version.

Some OEMs manufacturing JABWT devices may want to provide custom application environments for their devices. These manufacturers may want to extend JABWT in a proprietary way and provide additional functionality. This can be accomplished by defining LOCs or LCCs (see Chapter 1). But programs using these classes may not be portable across devices.

This book presents the need for JABWT, explains the overall architecture, and extensively discusses the various facets of JABWT—their use and programming techniques. The book gives insights into the intended use of the APIs. The book, we believe, gives enough code examples to help a programmer become proficient at programming with JABWT.

In summary, we believe the basic human desire to stay connected and communicate with computing devices from anywhere and at all times will increase the demand on wireless communications. Standard programming environments for accessing these wireless communications media will help create a myriad of applications. This book presents a simple yet powerful standard API for Bluetooth Wireless Technology. We hope the power of JABWT will encourage people to write more applications, write Bluetooth profiles with JABWT, and build more JABWT devices.

References

1. Bluetooth SIG. Specification of the Bluetooth System, Core v2.1, www.bluetooth.com, 2007.

2. Kumar, C B., P. J. Kline and T. J. Thompson. *Bluetooth Application Programming with the Java APIs*. San Francisco: Morgan Kaufmann, 2004.

3. Miller, B. A., and C. Bisdikian. *Bluetooth Revealed*, 2nd ed. Upper Saddle River: Prentice-Hall, 2001.

4. Bray, J. and C. F. Sturman. *Bluetooth 1.1: Connect without Cables*, 2nd ed. Upper Saddle River: Prentice Hall, 2001.

5. Bluetooth SIG, Bluetooth Network Encapsulation Protocol (BNEP) Specification, Revision 1.0, 2003.

6. Bluetooth SIG, Hardcopy Cable Replacement Profile Interoperability Specification, Revision 1.0a, 2002.

7. Bluetooth SIG, Audio/Video Control Transport Protocol Specification, Revision 1.2, 2007.

8. Bluetooth SIG, Audio/Video Distribution Transport Protocol Specification, Revision 1.2, 2007.

9. Bluetooth SIG, Serial Port Profile Specification, 2001.

10. Bluetooth SIG, Service Discovery Application Profile Specification, 2001.

11. Bluetooth SIG, Generic Object Exchange Profile Specification, 2001.

12. Bluetooth SIG, Bluetooth Qualification Program Website, www.bluetooth.com/Bluetooth/Apply/Technology/Qualification.

13. Topley, K. *J2ME in a Nutshell*. Sebastopol: O'Reilly, 2002.

14. Riggs, R., A. Taivalsaari, and M. Vandenbrink. *Programming Wireless Devices with the Java^{TM} 2 Platform, Micro Edition*. Boston: Addison-Wesley, 2001.

15. Java Community Process. J2ME Connected, Limited Devices Configuration (JSR-30), www.jcp.org/en/jsr/detail?id=30, 2000.

16. Java Community Process. J2ME Connected Device Configuration (JSR-36), www.jcp.org/en/jsr/detail?id=36, 2001.

17. Lindholm, T., and F. Yellin. *The Java^{TM} Virtual Machine Specification, Second Edition*. Boston: Addison-Wesley, 1999.

18. Java Community Process. Mobile Information Device Profile for the J2ME Platform (JSR-37), www.jcp.org/en/jsr/detail?id=37, 2000.

19. Java Community Process. J2ME Foundation Profile (JSR-46), www.jcp.org/en/jsr/detail?id=46, 2001.

20. Java Community Process. Personal Profile Specification (JSR-62), www.jcp.org/en/jsr/detail/?id=62, 2002.

21. Java Community Process. Personal Basis Profile Specification (JSR-129), www.jcp.org/en/jsr/detail/?id=129, 2002.

22. Java Community Process. Java APIs for Bluetooth Wireless Technology (JSR-82), www.jcp.org/en/jsr/detail/?id=82, 2002.

23. Java Community Process. Generic Connection Framework Optional Package for J2SE (JSR-197), www.jcp.org/en/jsr/detail/?id=197, 2003.

24. Kumar, C B., and P. Kline. "Bringing the Benefits of Java to Bluetooth." Embedded Systems, the European Magazine for Embedded Design (2002) no. 42.

25. Java Community Process. J2ME Connected, Limited Device Configuration 1.1 (JSR-139), www.jcp.org/en/jsr/detail/?id=139, 2003.

26. Java Community Process. Connected Device Configuration 1.1 (JSR-218), www.jcp.org/en/jsr/detail/?id=218, 2005.

27. Java Community Process. Mobile Information Device Profile 2.0 (JSR-118), www.jcp.org/en/jsr/detail/?id=118, 2002.

28. Java Community Process. Foundation Profile 1.1 (JSR-219), www.jcp.org/en/jsr/detail/?id=219, 2005.

29. Java Community Process. Personal Profile 1.1 (JSR-216), www.jcp.org/en/jsr/detail/?id=216, 2005.

30. Java Community Process. Personal Basis Profile 1.1 (JSR-217), www.jcp.org/en/jsr/detail/?id=217, 2005.

31. Bluetooth SIG. Basic Printing Profile, Revision 1.2, 2006.

32. Bluetooth SIG, Dial Up Networking Profile Specification, Revision 1.1, 2001.

33. Infrared Data Association®, IrDA® Object Exchange Protocol-OBEX™, version 1.3, 2003.

34. Bluetooth SIG. Bluetooth Assigned Numbers, www.bluetooth.org/technical/assignednumbers/home.htm.

35. Fowler, M., and K. Scott. *UML Distilled: A Brief Guide to the Standard Object Modeling Language*, 2nd ed. Boston: Addison-Wesley, 2000.

36. International Organization for Standardization. Code for the presentation of names of languages. ISO 639:1988 (E/F), Geneva, 1988.

37. Internet Assigned Numbers Authority. www.iana.org/assignments/character-sets, 2007.

38. The Internet Mail Consortium. vCard—The Electronic Business Card, Version 2.1, www.imc.org/pdi/, 1996.

39. The Open group. DCE 1.1: Remote Procedure Call, Appendix A. Document Number C706, www.opengroup.org/dce/info/faq-mauney.html#Q2_26, Reading, UK, 1997.

40. Bluetooth SIG, Personal Area Networking Profile, Revision 1.0, 2003.

41. Bluetooth SIG, Bluetooth Extended Service Discovery Profile (ESDP) for Universal Plug and Play™ (UPnP™), Revision 0.95a, 2003.

42. Bluetooth SIG, Audio/Video Remote Control Profile, Revision 1.3, 2007.

43. Bluetooth SIG, Generic Audio/Video Distribution Profile, Revision 1.2, 2007.

44. Bluetooth SIG, Advanced Audio Distribution Profile, Revision 1.2, 2007.

45. Bluetooth SIG, Hands-Free Profile, Revision 1.5, 2005.

46. Williams, S., and I. Millar. "The IrDA platform," in *Insights into Mobile Multimedia Communication*, ed. D. Bull, C. Canagarajah, and A. Nix. San Francisco: Morgan Kaufmann, 1998.

47. Bluetooth SIG, Headset Profile Specification, Revision 1.1, 2001.

Index